Functional Two-Dimensional Layered Materials—From Graphene to Topological Insulators

MATERIALS RESEARCH SOCIETY
SYMPOSIUM PROCEEDINGS VOLUME 1344

Functional Two-Dimensional Layered Materials—From Graphene to Topological Insulators

Symposium held April 25–29, 2011, San Francisco, California, U.S.A.

EDITORS

Alexander A. Balandin
University of California–Riverside,
Riverside, California, U.S.A.

Andre Geim
University of Manchester,
Manchester, United Kingdom

Jiaxing Huang
Northwestern University
Evanston, Illinois, U.S.A.

Dan Li
Monash University
Clayton, Victoria, Australia

Materials Research Society
Warrendale, Pennsylvania

CAMBRIDGE UNIVERSITY PRESS
Cambridge, New York, Melbourne, Madrid, Cape Town,
Singapore, São Paulo, Delhi, Mexico City

Cambridge University Press
32 Avenue of the Americas, New York, NY 10013-2473, USA

www.cambridge.org
Information on this title: www.cambridge.org/9781605113210

Materials Research Society
506 Keystone Drive, Warrendale, PA 15086, USA
http://www.mrs.org

First published 2012

CODEN: MRSPDH
ISBN: 978-1-60511-321-0 Hardback

CONTENTS

GRAPHENE AND GRAPHENE COMPOSITES: PROPERTIES AND APPLICATIONS

*Invited Paper

TOPOLOGICAL INSULATORS AND QUASI-2D MATERIALS

PREFACE

Two-dimensional (2D) layered materials attract growing interest in both fundamental science and technology. The most known 2D material – graphene – reveals excellent electronic, mechanical, and thermal properties. It was proposed for applications in transparent electrodes, sensing and advanced electron microscopy, energy conversion and storage, thermal management and high-frequency communication applications. Other 2D materials, quasi-2D crystals and molecular monolayers are also gaining visibility. One of the examples is quintuples or few-quintuple layers of bismuth telluride family of materials, which were identified as topological insulators. Topological insulators and graphene are related via Dirac type of electron dispersion.

Symposium Y, "Functional Two-Dimensional Layered Materials", at the 2011 MRS Spring Meeting in San Francisco, California, April 25-29 was the first MRS symposium which combined together graphene, topological insulator thin films, other 2D and quasi-2D materials aimed to highlight breakthroughs, progress, and challenges in the synthesis, processing, structure, and assembly of 2D layered materials, and how these factors affect their properties and applications. The session topics included: (i) materials synthesis and processing (physical and chemical approaches, top-down and bottom-up; control of size, shape, and conformation of the 2D sheets; processing and assembly, patterning and integration into device structures); (ii) structure and characterization (microscopy, spectroscopy, theory and simulation); (iii) physical and chemical properties (optical, electronic, thermal, magnetic, and mechanical properties; surface modification; chemical and photochemical reactivity); (iv) electronic, thermal management and energy applications (electron and phonon transport; thermal properties; composites, hybrid materials, catalytic, energy, and biological applications); (v) topological insulators and non-carbon 2D materials.

Among the invited speakers who took part in the symposium were such recognized experts in the field as Professor Pulickel Ajayan (Rice University), Professor Ilhan Aksay (Princeton University), Dr. Phaedon Avouris (IBM T.J. Watson Research Center), Professor Manish Chhowalla (Rutgers University), Professor Jonathan Coleman (Trinity College Dublin, Ireland), Professor Yi Cui (Stanford University), Dr. Andrea Ferrari (University of Cambridge, United Kingdom), Dr. Suchismita Ghosh (Intel Corporation), Professor Robert Haddon (University of California, Riverside), Professor Mark Hersam (Northwestern University), Professor Richard Kaner (University of California, Los Angeles), Professor Philip Kim (Columbia University), Professor Sang Ouk Kim (Korea Advanced Institute of Science and Technology), Professor Roger Lake (University of California, Riverside), Dr. Jun Liu (Pacific Northwest National Laboratory), Professor Kian-Ping Loh (National University of Singapore), Professor Son Binh Nguyen (Northwestern University), Professor Elena Obraztsova (General Physics Institute, Russian Academy of Sciences, Moscow), Professor Rodney Ruoff (University of Texas, Austin), Professor James Tour (Rice University), Professor Jinlong Yang (University of Science and Technology of China), Professor Alex Zettl (University of California, Berkeley), and Professor Hua Zhang (Nanyang Technological University, Singapore).

Symposium Y was attended by a large number of graduate students who gave talks and presented posters. Some of their outstanding work and presentations have been recognized by the MRS awards. For example, an electrical engineering PhD candidate Desalegne Teweldebrhan, who conducts his research in Professor Balandin's Nano-Device Laboratory at the University of California – Riverside, received the MRS Graduate Student Silver Award for his work on tuning of graphene properties via controlled exposure to electron beam irradiation. Jaemyung Kim, a PhD candidate from Professor Jiaxing Huang's group in the Materials Science and Engineering Department at Northwestern University also received the MRS Graduate Student Silver Award for developing the fluorescence quenching microscopy (FQM) technique for seeing graphene-based sheets, and revealing the surfactant-like behaviors of graphene oxide. In addition, Dr. Hisato Yamaguchi from Professor Manish Chhowalla' group in the Department of Materials Science and Engineering at Rutgers, The State University of New Jersey, received a Best Poster Award for his poster "Field Emission from Atomically Thin Edges of Reduced Graphene Oxide".

Symposium Y also created Symposium Young Investigator Awards to recognize the outstanding presentations given by students and postdocs. The winners were determined by voting of several symposium organizers and invited speakers. The winners of the Symposium's first place awards were Guangyu Xu, Laura J. Cote, Long Ju, Guanxiong Liu, and Vincent C. Tung. The second place award recipients were Shu Nie, Craig M. Nolen, Yu-Ying Lee, Tae Hee Han, and Owen Compton. The third place awards went to Zhong Yan, Zahid Hossain, Jie Yu, Shirui Guo, Shaahin Amini, Javed Khan, and Jiayan Luo. Other students and postdocs received a copy of the DVD "NOVA: Making Stuff" as a souvenir. The Symposium Y awards and souvenirs were sponsored by Materials Today – Elsevier, Princeton Instruments, and Cambridge NanoTech.

This volume presents a selection of papers, presented at the MRS Symposium Y, "Functional Two-Dimensional Layered Materials", which were submitted for publication. It provides an overview of the research topics and possible applications of the 2D materials and related systems. A number of the Symposium Y award winning papers, presented by the graduate students and postdoctoral researchers, are among those included in the volume. We hope that the volume will be interesting and stimulating for a wide audience.

Alexander A. Balandin
Andre Geim
Jiaxing Huang
Dan Li

August 2011

ACKNOWLEDGMENTS

The papers published in this volume result from Symposium Y, "Functional Two-Dimensional Layered Materials", at the 2011 MRS Spring Meeting in San Francisco, California. We sincerely thank all of the oral and poster presenters of the symposia who contributed to this proceedings volume. We also thank the reviewers of these manuscripts, who provided valuable feedback to the editors and to the authors. It is an understatement to say that the symposia and the proceedings would not have happened without the organizational help of the Materials Research Society and its staff.
The organizers of Symposium Y thank Materials Today – Elsevier Ltd., Princeton Instruments, and Cambridge NanoTech Inc. for their financial support.

MATERIALS RESEARCH SOCIETY SYMPOSIUM PROCEEDINGS

Volume 1321 — Amorphous and Polycrystalline Thin-Film Silicon Science and Technology—2011, B. Yan, S. Higashi, C.C. Tsai, Q. Wang, H. Gleskova, 2011, ISBN 978-1-60511-298-5
Volume 1322 — Third-Generation and Emerging Solar-Cell Technologies, Y. Lu, J.M. Merrill, M.T. Lusk, S. Bailey, A. Franceschetti, 2011, ISBN 978-1-60511-299-2
Volume 1323 — Advanced Materials Processing for Scalable Solar-Cell Manufacturing, H. Ji, B. Ren, V. Mannivanan, L. Tsakalakos, 2011, ISBN 978-1-60511-300-5
Volume 1324 — Compound Semiconductors for Energy Applications and Environmental Sustainability—2011, F. Shahedipour-Sandvik, L.D. Bell, K. Jones, B. Simpkins, D. Schaadt, M. Contreras, 2011, ISBN 978-1-60511-301-2
Volume 1325 — Energy Harvesting—Recent Advances in Materials, Devices and Applications, R. Venkatasubramanian, H. Liang, H. Radousky, J. Poon, 2011, ISBN 978-1-60511-302-9
Volume 1326E — Renewable Fuels and Nanotechnology, H. Idriss, 2011, ISBN 978-1-60511-303-6
Volume 1327 — Complex Oxide Materials for Emerging Energy Technologies, J.D. Perkins, A. Ohtomo, H.N. Lee, G. Herranz, 2011, ISBN 978-1-60511-304-3
Volume 1328E — Electrochromic Materials and Devices, M. Bendikov, D. Gillaspie, T. Richardson, 2011, ISBN 978-1-60511-305-0
Volume 1329E — Nanoscale Heat Transfer—Thermoelectrics, Thermophotovoltaics and Emerging Thermal Devices, K. Nielsch, S.F. Fischer, B.J.H. Stadler, T. Kamins, 2011, ISBN 978-1-60511-306-7
Volume 1330E — Protons in Solids, V. Peterson, 2011, ISBN 978-1-60511-307-4
Volume 1331E — Frontiers of Solid-State Ionics, K. Kita, 2011, ISBN 978-1-60511-308-1
Volume 1332E — Interfacial Phenomena and *In-Situ* Techniques for Electrochemical Energy Storage and Conversion, H. Li, 2011, ISBN 978-1-60511-309-8
Volume 1333E — Nanostructured Materials for Energy Storage, J. Lemmon, 2011, ISBN 978-1-60511-310-4
Volume 1334E — Recent Developments in Materials for Hydrogen Storage and Carbon-Capture Technologies, R. Zidan, 2011, ISBN 978-1-60511-311-1
Volume 1335 — Materials, Processes, and Reliability for Advanced Interconnects for Micro- and Nanoelectronics—2011, M. Baklanov, G. Dubois, C. Dussarrat, T. Kokubo, S. Ogawa, 2011, ISBN 978-1-60511-312-8
Volume 1336E — Interface Engineering for Post-CMOS Emerging Channel Materials, Y. Kamata, 2011, ISBN 978-1-60511-313-5
Volume 1337 — New Functional Materials and Emerging Device Architectures for Nonvolatile Memories, D. Wouters, O. Auciello, P. Dimitrakis, Y. Fujisaki, E. Tokumitsu, 2011, ISBN 978-1-60511-314-2
Volume 1338E — Phase-Change Materials for Memory and Reconfigurable Electronics Applications, B.-K. Cheong, P. Fons, B.J. Kooi, B.-S. Lee, R. Zhao, 2011, ISBN 978-1-60511-315-9
Volume 1339E — Plasma-Assisted Materials Processing and Synthesis, J.L. Endrino, A. Anders, J. Andersson, D. Horwat, M. Vinnichenko, 2011, ISBN 978-1-60511-316-6
Volume 1340E — High-Speed and Large-Area Printing of Micro/Nanostructures and Devices, T. Sekitani, 2011, ISBN 978-1-60511-317-3
Volume 1341 — Nuclear Radiation Detection Materials—2011, A. Burger, M. Fiederle, L. Franks, K. Lynn, D.L. Perry, K. Yasuda, 2011, ISBN 978-1-60511-318-0
Volume 1342 — Rare-Earth Doping of Advanced Materials for Photonic Applications—2011, V. Dierolf, Y. Fujiwara, T. Gregorkiewicz, W.M. Jadwisienczak, 2011, ISBN 978-1-60511-319-7
Volume 1343E — Recent Progress in Metamaterials and Plasmonics, G.J. Brown, J. Pendry, D. Smith, Y. Lu, N.X. Fang, 2011, ISBN 978-1-60511-320-3
Volume 1344 — Functional Two-Dimensional Layered Materials—From Graphene to Topological Insulators, A.A. Balandin, A. Geim, J. Huang, D. Li, 2011, ISBN 978-1-60511-321-0
Volume 1345E — Nanoscale Electromechanics of Inorganic, Macromolecular and Biological Systems, J. Li, S.V. Kalinin, M.-F. Yu, P.S. Weiss, 2011, ISBN 978-1-60511-322-7
Volume 1346 — Micro- and Nanofluidic Systems for Materials Synthesis, Device Assembly and Bioanalysis—2011, R. Fan, J. Fu, J. Qin, A. Radenovic, 2011, ISBN 978-1-60511-323-4
Volume 1347E — Nanoscale Heat Transport—From Fundamentals to Devices, A. McGaughey, M. Su, S. Putnam, J. Shiomi, 2011, ISBN 978-1-60511-324-1

Materials Research Society Symposium Proceedings

Volume 1348E — Hybrid Interfaces and Devices, D.S. Ginley, N.R. Armstrong, G. Frey, R.T. Collins, 2011, ISBN 978-1-60511-325-8
Volume 1349E — Quantitative Characterization of Nanostructured Materials, A. Kirkland, 2011, ISBN 978-1-60511-326-5
Volume 1350E — Semiconductor Nanowires—From Fundamentals to Applications, V. Schmidt, L.J. Lauhon, T. Fukui, G.T. Wang, M. Björk, 2011, ISBN 978-1-60511-327-2
Volume 1351 — Surfaces and Nanomaterials for Catalysis through In-Situ or Ex-Situ Studies, F. Tao, M. Salmeron, J.A. Rodriguez, J. Hu, 2011, ISBN 978-1-60511-328-9
Volume 1352 — Titanium Dioxide Nanomaterials, X. Chen, M. Graetzel, C. Li, P.D. Cozzoli, 2011, ISBN 978-1-60511-329-6
Volume 1353E — The Business of Nanotechnology III, L. Merhari, M. Biberger, D. Cruikshank, M. Theelen, 2011, ISBN 978-1-60511-330-2
Volume 1354 — Ion Beams—New Applications from Mesoscale to Nanoscale, G. Marletta, A. Öztarhan, J. Baglin, D. Ila, 2011, ISBN 978-1-60511-331-9
Volume 1355E — Biological Hybrid Materials for Life Sciences, L. Stanciu, S. Andreescu, T. Noguer, B. Liu, 2011, ISBN 978-1-60511-332-6
Volume 1356E — Microbial Life on Surfaces—Biofilm-Material Interactions, N. Abu-Lail, W. Goodson, B.H. Lower, M. Fornalik, R. Lins, 2011, ISBN 978-1-60511-333-3
Volume 1357E — Biomimetic Engineering of Micro- and Nanoparticles, D. Discher, 2011, ISBN 978-1-60511-334-0
Volume 1358E — Organic Bioelectronics and Photonics for Sensing and Regulation, L. Torsi, 2011, ISBN 978-1-60511-335-7
Volume 1359 — Electronic Organic and Inorganic Hybrid Nanomaterials—Synthesis, Device Physics and Their Applications, Z.-L. Zhou, C. Sanchez, M. Popall, J. Pei, 2011, ISBN 978-1-60511-336-4
Volume 1360 — Synthesis and Processing of Organic and Polymeric Materials for Semiconductor Applications, A.B. Padmaperuma, J. Li, C.-C. Wu, J.-B. Xu, N.S. Radu, 2011, ISBN 978-1-60511-337-1
Volume 1361E — Engineering Polymers for Stem-Cell-Fate Regulation and Regenerative Medicine, S. Heilshorn, J.C. Liu, S. Lyu, W. Shen, 2011, ISBN 978-1-60511-338-8
Volume 1362 — Carbon Functional Interfaces, J.A. Garrido, K. Haenen, D. Ho, K.P. Loh, 2011, ISBN 978-1-60511-339-5
Volume 1363 — Fundamental Science of Defects and Microstructure in Advanced Materials for Energy, B.P. Uberuaga, A. El-Azab, G.M. Stocks, 2011, ISBN 978-1-60511-340-1
Volume 1364E — Forum on Materials Education and Evaluation—K-12, Undergraduate, Graduate and Informal, B.M. Olds, D. Steinberg, A. Risbud, 2011, ISBN 978-1-60511-341-8
Volume 1365 — Laser-Material Interactions at Micro/Nanoscales, Y. Lu, C.B. Arnold, C.P. Grigoropoulos, M. Stuke, S.M. Yalisove, 2011, ISBN 978-1-60511-342-5
Volume 1366E — Crystalline Nanoporous Framework Materials—Applications and Technological Feasibility, M.D. Allendorf, K. McCoy, A.A. Talin, S. Kaskel, 2011, ISBN 978-1-60511-343-2
Volume 1367E — Future Directions in High-Temperature Superconductivity—New Materials and Applications, C. Cantoni, A. Ballarino, K. Matsumoto, V. Solovyov, H. Wang, 2011, ISBN 978-1-60511-344-9
Volume 1368 — Multiferroic, Ferroelectric, and Functional Materials, Interfaces and Heterostructures, P. Paruch, E. Tsymbal, M. Rzchowski, T. Tybell, 2011, ISBN 978-1-60511-345-6
Volume 1369E — Computational Studies of Phase Stability and Microstructure Evolution, Z.-K. Liu, 2011, ISBN 978-1-60511-346-3
Volume 1370 — Computational Semiconductor Materials Science, L. Liu, S.-H. Wei, A. Rubio, H. Guo, 2011, ISBN 978-1-60511-347-0

Prior Materials Research Society Symposium Proceedings available by contacting Materials Research Society

Graphene and Graphene Composites: Synthesis, Processing and Characterization

Mater. Res. Soc. Symp. Proc. Vol. 1344 © 2011 Materials Research Society
DOI: 10.1557/opl.2011.1366

Langmuir-Blodgett Assembly of Soft Carbon Sheets

Laura J. Cote, Jaemyung Kim and Jiaxing Huang
Department of Materials Science and Engineering, Northwestern University, Evanston, USA

Abstract

Graphene oxide sheets have recently gained immense interest as a building block for graphene based materials and devices. Rapid developments have been made in the chemistry and applications of GO. However, assembly, too, plays a critical role in the final properties of bulk graphene based materials as it determines the microstructures of the 2D sheets. There is thus a pressing need for controllable assembly strategies. Based on the recent identification of the pH dependent surfactant-like behavior of GO sheets, we are now able to control the tiling morphologies of such sheets to produce thin films with either wrinkled or overlapped types of microstructures. This allows for the deconvolution of the effects of these two basic morphological features in the electrical and optical properties of the resulting thin films, providing a well-defined example of the processing-microstructure-properties relationship for this unique soft material building block.

Introduction

Graphite oxide sheets, now known as graphene oxide (GO), are the product of the liquid-phase oxidation-exfoliation reaction of graphite powder with strong oxidants such as a mixture of sulfuric acid and potassium permanganate.[1,2] The resultant GO consists of exfoliated single atomic layers of sp^2-hybridized carbon atoms derivatized by phenol hydroxyl and epoxide groups across the basal plane and ionizable carboxylic acid groups along the edges,[3,4] as shown schematically in Fig. 1a. Charge repulsion of the ionized acid groups allows GO to form stable single layer aqueous dispersions.[5] In recent years, rapid progress has been made in the chemistry[6-8] and applications of GO as a solution-processable precursor for graphene-based materials and devices.[9-12] However, there remains a pressing need for rational assembly strategies of these two-dimensional (2D) sheets since assembly plays a critical role in controlling microstructures and thus properties of the final materials.

The combination of the hydrophilic hydroxyl and carboxyl acid groups on the edges with the unoxidized hydrophobic polyaromatic nanographene domains remaining on the basal plane (Fig. 1a) also makes GO a unique tethered 2D surfactant sheet. Indeed, GO has been found to be enriched at interfaces without being modified by surfactants[13] and capable of lowering interfacial energies.[14] Additionally, the amphiphilicity of GO can be tuned by pH since it affects the degree of ionization (Fig. 1b) or number of charges on the sheets, as confirmed by zeta potential measurements (Fig. 1c).

The surfactant-like properties of GO suggest that interfacial assembly of the sheets could potentially be controlled by tuning their amphiphilicity. Indeed, pH can be used as a tuning parameter to guide the assembly of GO sheets at the air-water interface. Interacting sheets can be coaxed into a continuous film with two distinct types of microscopic morphologies, namely wrinkles and partial overlaps. The microstructures were found to have different effects on the electrical and optical properties of the final graphene-based monolayers. These processing-microstructure-properties relationships can provide a model system for designing improved graphene-based technologies such as thin film transparent conductors[15-21] since wrinkles and overlaps are the two fundamental morphological features of graphene based thin films.

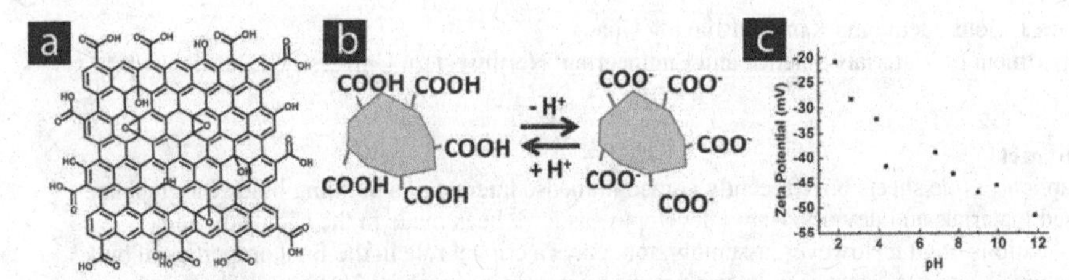

Fig. 1 GO sheets can act as a 2D surfactant with amphiphilicity tunable by pH. (a) The structural model of GO shows hydrophobic polyaromatic domains in its basal plane and hydrophilic carboxylic acid groups along the edge. (b) The degree of ionization of GO can be reversibly tuned by pH, which is confirmed by (c) zeta potential measurements.

Interfacial Assembly of GO Sheets

Since the surfactant nature of GO allows the sheets to float on a water surface without the need for stabilizing agents, classical molecular assembly methods such as the Langmuir-Blodgett (LB) technique can be used to create large-area monolayers. In the classical LB technique, amphiphilic molecules are first dissolved in a volatile organic solvent and then spread dropwise on the water surface and confined by movable barriers. As the solvent evaporates, the molecules are trapped at the air-water interface to form a monolayer. As the barriers are closed, the surface density of the molecules increases, leading to an increase in surface pressure, or reduction in surface tension which can be monitored using a tensiometer. The floating monolayers can be transferred to a solid substrate by vertical dip-coating. The two-dimensional water surface should serve as an ideal platform to assemble GO sheets. First, the interface is geometrically similar to GO, making it ideal to accommodate the flat sheets; second, the soft, fluidic "substrate" should allow free movement of GO sheets upon manipulation, which should facilitate edge-to-edge or face-to-face interactions of the flat GO sheets.

When droplets of an optimal 5:1 methanol:water GO dispersion are gently deposited onto the water surface, they can first spread rapidly on the surface before mixing with water.[13] In this way, the GO surfactant sheets can be effectively trapped at the air-water interface. The density of the sheets can be continuously tuned by moving the barriers. Upon compression, the monolayer exhibits a gradual increase in surface pressure, as shown in the surface-pressure–area isotherm plot in Fig. 2e. Films collected by dip-coating along the compression curve were imaged by scanning electron microscopy (SEM) and showed four distinct packing behaviors. Initially, where the surface pressure is near zero, the collected film consists of dilute, well-isolated flat sheets (Fig. 2a). As compression continues, a gradual increase in surface pressure begins to occur

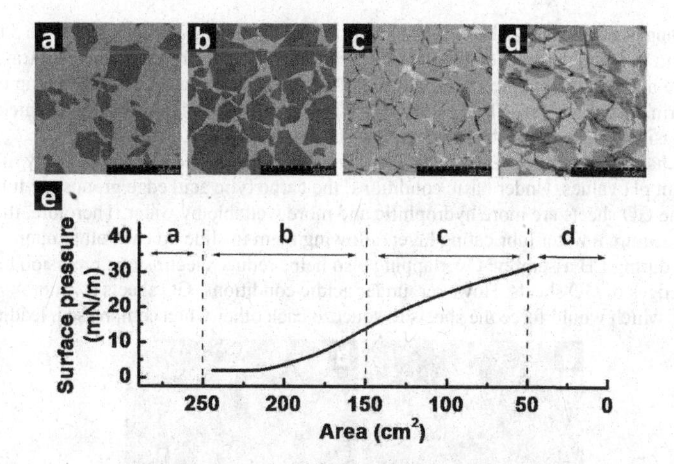

Fig. 2 Langmuir-Blodgett assembly of single layer graphite oxide sheets. (a-d) SEM images showing the collected GO monolayers on a silicon wafer at different regions in the isotherm. The packing density was continuously tuned: (a) dilute monolayer of isolated flat sheets, (b) monolayer of close-packed GO, (c) over-packed monolayer with sheets folded at interconnected edges, and (d) over-packed monolayer with folded and partially overlapped interlocking sheets. (e) Isothermal surface pressure-area plot showing corresponding (a-d) collection regions. Scale bars represent 20 µm.

and the sheets start to close pack into a broken tile mosaic pattern over the entire surface (Fig. 2b). Upon further compression, the soft sheets are forced to fold and wrinkle at their contact points in order to accommodate the increased pressure (Fig. 2c). Even further compression resulted in interlocked sheets with nearly complete surface coverage (Fig. 2d). LB assembly can produce flat GO thin films with uniform and continuously tunable coverage, thus avoiding the uncontrollable wrinkles, overlaps, and voids in films fabricated by other techniques.

Structure-Property Relationships

The hydrophilicity of GO can be tuned by the degree of ionization of the carboxylic acid groups through pH, which directly affects their surface activity. To investigate the pH dependent assembly of GO at the air-water interface, we used the LB technique to produce GO monolayers under both acidic and basic conditions. These monolayers were collected at the end of compression and characterized using SEM[13, 22] and atomic force microscopy (AFM). SEM images revealed many randomly oriented smaller wrinkles on individual GO sheets in the "acidic" monolayer (Fig. 3a). Since the GO sheets are irregularly shaped polyhedrons with polydisperse sizes, they squeeze each other from random directions when compressed together, producing these disoriented wrinkles. Additionally, many wrinkles near the edges of the GO sheets were aligned with the contact lines of the neighboring sheets, suggesting they were also buckling structures developed under in-plane compressive forces. AFM images on individual wrinkles show that they are typically thicker than 3 nm, often reaching 5 to 10 nm (Fig. 3b).

Because the apparent thickness of GO is around 1 nm, these wrinkles effectively create 5 to 10-layer islands in the monolayer, thereby increasing the film's surface roughness. In contrast, wrinkles were not observed in the "basic" film. Rather, the sheets seemed to slide on top of one another to form partial overlaps. These overlapped areas were typically only 2 layers thick, making the basic monolayer much smoother than the acidic one (Fig. 3b,d).

This change in packing behavior is likely due to the difference in hydrophilicity of GO under different pH values. Under basic conditions, the carboxylic acid edge groups are fully ionized, so the GO sheets are more hydrophilic and more wettable by water. Therefore, the sheets could acquire a water lubricating layer, allowing them to slide on each other upon compression during LB assembly. Overlapping also helps reduce electrostatic repulsion between the charged edges of GO sheets. However, under acidic conditions, GO sheets are more hydrophobic, which would force the sheets to squeeze each other when compressed, leading to

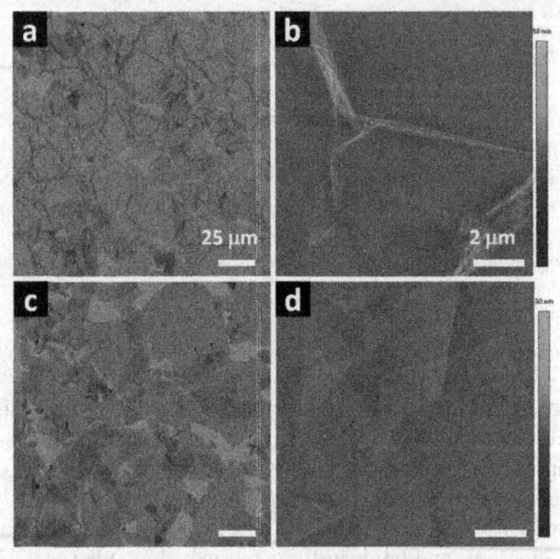

Fig. 3 Wrinkles and overlaps: pH dependent packing of GO sheets in the LB monolayer. The morphology of interacting GO sheets can be tuned by changing the pH values of the subphase as shown by the SEM and AFM images (a,c, SEM, scale bar 25 μm and b,d, AFM, scale bar 2 μm. Upon compression, GO sheets tend to wrinkle (a,b) on acidic subphase and overlap (c,d) on basic subphase.

buckling structures. Hydrogen bonding between the protonated edge carboxylic acid groups may also prevent the sheets from sliding, forcing wrinkles to form during compression.

Wrinkles and overlaps are the two fundamental morphologies of interacting sheets in GO based bulk materials. pH dependent LB assembly can offer model systems to investigate the effects of these microstructures on materials properties. Here, we present a study on how wrinkles and overlaps affect sheet resistance and optical transmission.

To study sheet resistance, the GO monolayers were thermally reduced, and gold electrodes were patterned on the resulting r-GO films by evaporation. Since the r-GO films were made by tiled smaller pieces, the distribution of sheet resistance values measured at different

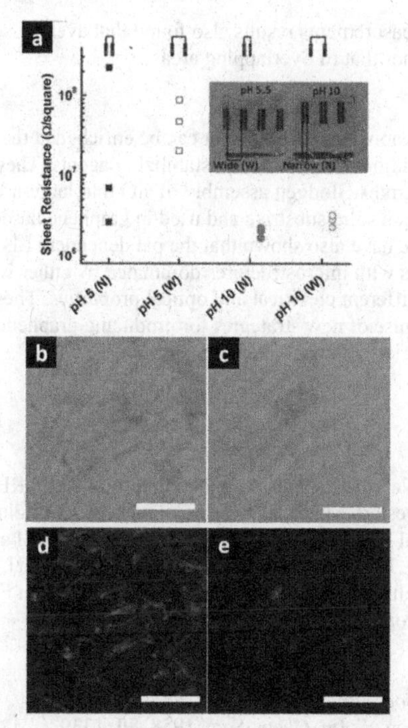

Fig. 4 Effect of pH on sheet resistance and optical transmission. (a) Sheet resistances of r-GO monolayers collected from a pH 5 (black squares) and pH 10 (red circles) subphases. Inset shows samples with patterned gold electrodes of wide and narrow separations of 2.5 and 0.25 nm, respectively. For the "acidic" monolayer, the sheet resistance values measured at different spots and electrode separations were very scattered, suggesting uneven coverage of GO sheets. In contrast, the resistance values for the "basic" monolayer were much more uniform and consistent., suggesting uniform coverage. Optical microscopy under bright-field transmission mode (b,c) and dark field scattering mode (d,e) reveal that wrinkles (b,d) cause higher optical loss than overlaps (c,e).

spots and electrode separations reflects uniformity of sample coverage. The sheet resistance values measured for the acidic monolayer have a very wide distribution, especially with narrower electrode separation. In contrast, values obtained with the basic monolayer were much more focused and consistent regardless of electrode separation. Overall, sheet resistance values of the acidic r-GO monolayers were only slightly higher but much more scattered than those of the basic monolayers, indicating a much greater degree of non-uniformity in the acidic monolayers.

The optical properties of the wrinkled and overlapped samples also differ. Optical loss is greater in the wrinkled structures, which have an average percent transmission of 94.1% compared to 96.1% for the partially overlapped, hydrazine reduced GO monolayers. Bright field transmission optical microscopy images of a reduced wrinkled film show darker features (i.e. increased loss) at the wrinkled sites. The cause of the increased loss is revealed in dark field images, where wrinkles can be clearly seen as bright scattering centers on the black background. In contrast, the overlaps are essentially featureless in both bright field and dark field modes of imaging. Local laser scattering measurements results also found that average scattering from wrinkled areas was around 3.7 times that of overlapping areas.

Conclusion

GO is a unique two-dimensional amphiphile that can be enriched at the air-water interface without need for any additional surfactants or stabilizing agents. These surfactant properties allow the scalable Langmuir-Blodgett assembly of GO into large-area monolayers which can be easily deposited onto a solid substrate and used in graphene based materials upon thermal or chemical reduction. We have also shown that the pH dependent LB assembly of GO can be used to prepare monolayers with microstructures dominated by either wrinkles or overlaps, resulting in films with different electrical and optical properties. These studies on the LB assembly of GO hold the promise of new strategies for producing graphene based thin films with well-tuned properties in a variety of novel applications.

Acknowledgements

This work was supported by the National Science Foundation through a CAREER award (DMR 0955612). J.H. thanks the Northwestern Materials Research Science and Engineering Center (NSF DMR-0520513) for a capital equipment fund for the purchase of BAM and additional support from Sony Corporation. L.J.C. is a NSF graduate research fellow. J.H. is an Alfred P. Sloan Research Fellow. J.K. gratefully acknowledges support from the Ryan Fellowship and the Northwestern University International Institute for Nanotechnology.

Notes and References

1 B.C. Brodie, *Philos. Trans. R. Soc. London*, 1859, **149**, 249-259.
2 W.S. Hummers and R.E. Offeman, *J. Am. Chem. Soc.*, 1958, **80**, 1339.
3 W. Cai, R.D. Piner, F.J. Stadermann, S. Park, M.A. Shaibat, Y. Ishii, D. Yang, A. Velamakanni, S.J. An, M. Stoller, J. An, D. Chen and R.S. Ruoff, *Science*, 2008, **321**, 1815-1817.
4 A. Lerf, H.Y. He, M. Forster and J. Klinowski, *J. Phys. Chem. B*, 1998, **102**, 4477-4482.
5 D. Li, M.B. Muller, S. Gilje, R.B. Kaner, and G.G. Wallace, *Nat. Nanotechnol.*, 2008, **3**, 101-105.
6 W. Gao, L.B. Alemany, L. Ci, and P.M. Ajayan, *Nat. Chem.*, 2009, **1**, 403-408.
7 A. Bagri, C. Mattevi, M. Acik, Y.J. Chabal, M. Chhowalla and V.B. Shenoy, *Nat. Chem.*, 2010, **2**, 581-587.
8 Z. Wei, D. Wang, S. Kim, S.Y. Kim, Y. Hu, M.K. Yakes, A.R. Laracuente, Z. Dai, S.R. Marder, C. Berger, W.P. King, W.A. de Heer, P.E. Sheehan and E. Riedo, *Science*, 2010, **328**, 1373-1376.
9 D. Li and R.B. Kaner, *Science*, 2008, **320**, 1170-1171.
10 M.J. Allen, V.C. Tung and R.B. Kaner, *Chem. Rev.*, 2010, **110**, 132-145.
11 S. Park and R.S. Ruoff, *Nat. Nanotechnol.*, 2009, **4**, 217-224.
12 O.C. Compton and S.T. Nguyen, *Small*, 2010, **6**, 711-723.
13 L.J. Cote, F. Kim, and J. Huang, *J. Am. Chem. Soc.*, 2009, **131**, 1043-1049.
14 J. Kim, L.J. Cote, F. Kim, W. Yuang, K.R. Shull and J. Huang, *J. Am. Chem. Soc.*, 2010, **132**, 8180-8186
15 X. Wang, L. Zhi and K. Mullen, *Nano Lett.*, 2007, **8**, 323-327.

16 G. Eda, G. Fanchini and M. Chhowalla, *Nat. Nanotechnol.*, 2008, **3**, 270-274.
17 V.C. Tung, M.J. Allen, Y. Yang, and R.B. Kaner, *Nat. Nanotechnol.*, 2009, **4**, 25-29.
18 H.A. Becerril, J. Mao, Z. Liu, R.M. Stoltenberg, Z. Bao and Y. Chen, *ACS Nano*, 2008, **2**, 463-470.
19 S. De and J.N. Coleman, *ACS Nano*, 2010, **4**, 2713-2720.
20 Y.Y. Liang, J. Frisch, L.J. Zhi, H. Norouzi-Arasi, X.L. Feng, J.P. Rabe. N. Koch and K. Mullen, *Nanotechnology*, 2009, **20**, 4007.
21 Q. Su, S.P. Pang, V. Alijani, C. Li, X.L. Feng and K. Mullen, *Adv. Mater.*, 2009, **21**, 3191-3195.
22 J. Kim, F. Kim and J. Huang, *Mater. Today*, 2010, **13**, 28-38.

Mater. Res. Soc. Symp. Proc. Vol. 1344 © 2011 Materials Research Society
DOI: 10.1557/opl.2011.1354

Reversible Tuning of the Electronic Properties of Graphene via Controlled Exposure to Electron Beam Irradiation and Annealing

Desalegne Teweldebrhan[1, 2], Guanxiong Liu[1], and Alexander A. Balandin[1, 2]
[1]Nano-Device Laboratory, Bourne College of Engineering, University of California – Riverside, Riverside, CA 92521 U.S.A.
[2]Material Science and Engineering Program, University of California – Riverside, Riverside, CA 92521 U.S.A.

ABSTRACT

Graphene reveals many extraordinary properties including extremely high room temperature carrier mobility and intrinsic thermal conductivity. Understanding how to controllably modify graphene's properties is essential for its proposed applications. Here we report on a method for tuning the electrical properties of graphene via electron beam irradiation. It was observed that single-layer graphene is highly susceptible to the low-energy electron beams. We demonstrated that by controlling the irradiation dose one can change, by desired amount, the carrier mobility, shift the charge neutrality point, increase the resistance at the minimum conduction point, induce the "transport gap" and achieve current saturation in graphene. The change in graphene properties is due to defect formation on the graphene surface and in the graphene lattice. The changes are reversible by annealing until some critical irradiation dose is reached.

INTRODUCTION

Graphene is a single planer sheet of sp^2-bound carbon atoms with many extraordinary properties. Such unique properties include an extremely high room temperature (RT) carrier mobility of up to ~15,000 cm^2/Vs [1-2] and an extremely high intrinsic thermal conductivity exceeding ~3000 W/mK near RT for large flakes [3-4]. Recent experiments with modification of graphene surface via hydrogenation [5-6], ions irradiation [7], fluorination [8], and adsorption of individual gas molecules (NO_2, NH_3, etc.) [7] have shown that graphene's properties can be altered and tuned for specific applications. However, little is known about the effect of the electron beam irradiation on graphene or graphene-based devices. Recently, it was also shown that graphene exposure to the electron beams (e-beams) results in modification of its surface [9-11]. It was demonstrated that electron irradiation leads to the appearance of the disorder D peak at ~1350 cm^{-1} in the Raman spectra of irradiated graphene [9]. Here we describe how the electrical properties of the single layer graphene (SLG) depend on the irradiation dose, and correlate the current – voltage characteristics with the evolution of Raman spectrum of irradiated graphene.

EXPERIMENTAL DETAILS

Graphene back-gate FET devices were fabricated with the electron beam lithography (EBL).

Degenerately p+ doped silicon substrate was used to back-gate and tune the Fermi level in graphene. The graphene flakes were prepared by the standard micro-mechanical exfoliation from the high quality graphite. The flakes were transferred to the silicon substrate with 300-nm-thick layer of silicon oxide. Raman spectroscopy was used to verify the number of layers and check their quality. The details of our Raman inspection procedures were reported by us elsewhere [12-13]. The SLG samples were selected via de-convolution of the Raman 2D band and comparison of the intensities of the G peak and 2D band. A Reinshaw InVia micro-Raman spectrometer system was used with the laser wavelength of 488 nm. E-beam irradiation was conducted using a Leo SUPRA 55 system, which allows for accurate control of the exposed area and irradiation dose. The area dosage was calculated and controlled by the nanometer pattern generation system (NPGS). We selected the accelerating voltage of 20 keV and a working distance of 6 mm. The beam current for all the irradiation experiments on these devices was 30.8 pA. The graphene back gate devices were fabricated with the electron beam lithography (EBL). The electrical measurements were performed with an Agilent 4142B instrument.

RESULTS AND DISCUSSION

We carefully examined the Raman spectrum of the graphene devices after each irradiation step. For pristine graphene Raman has typical signatures of SLG: symmetric and sharp 2D band (~2700 cm^{-1}) and large I(2D)/I(G) ratio. The absent or undetectably small D peak at 1350 cm^{-1} indicates the defect-free high-quality graphene. The disorder D peak appears after the electron beam irradiation. Initially the intensity of the D grows with increasing dosage after each irradiation step. The trend reverses after the irradiation dose reaches a certain level. We used the intensity ratio I(D)/I(G) to characterize the relative strength of the D peak [9-11]. The ratio I(D)/I(G) reveals a clear and reproducible non-monotonic dependence on the irradiation dose . A similar trend was reported for graphite, where such dependence was attributed to the crystal structure change from crystalline to nanocrystalline and then to amorphous form [14]. The bond breaking in such cases is likely chemically induced since the electron energy is not sufficient for the ballistic knock out of the carbon atoms [9].

Our electrical measurements are consistent with this interpretation indicating a growing density of the charged impurities with increasing irradiation dose (see inset to Fig. 1b). Fig. 1b shows the evolution of the resistivity near the charge neutrality point with the irradiation dose. One can see a clear trend of increasing ρ_{max} with the irradiation dose. Since the contacts were not irradiated during the experiment, the overall increase of device resistance is due to the increasing resistivity of the irradiated graphene. This can be understood by the induced defects that create an increasing number of short range scattering centers in the graphene lattice. Note that the ρ_{max} increases by a factor of ~ 3 to 7 for SLG devices.

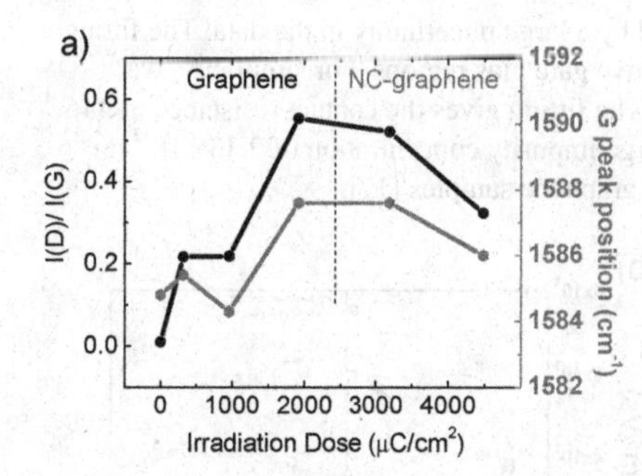

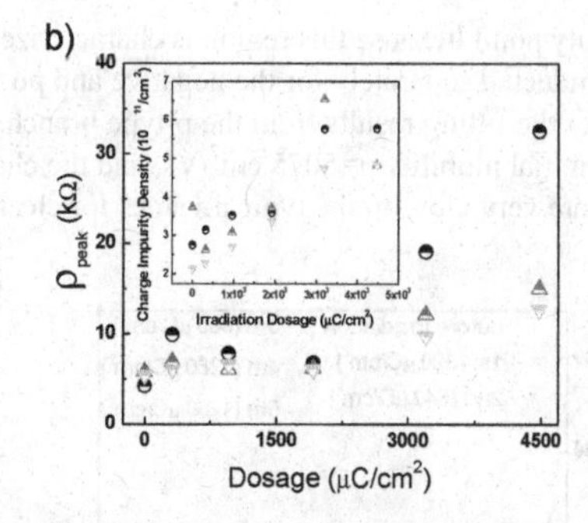

Figure 1: (a) Evolution of Raman spectrum of SLG with increasing irradiation dose. The initial defect-free graphene is irradiated by electron beam. The irradiation results in a prominent D (~1350 cm^{-1}) peak. (b) Evolution of SLG resistivity with increasing irradiation dose. Inset shows the effect of e-beam irradiation on the charge impurity density for SLG.

In order to analyze the results and rule out the role of the contact resistance we used the following equation to fit our resistance data [15, 16]

$$R_{DS} = R_{Cont} + \frac{L}{W}\left(\frac{1}{e\mu(\sqrt{n_0^2 + n_{BG}^2})}\right) + \frac{L}{W}\rho_{irr} \quad (1)$$

$$n_{BG} = \frac{C_{BG}\left|V_{BG} - V_{BG,\min}\right|}{e}, \quad (2)$$

where R_{Cont} is the contact resistance, μ is the mobility, e is the elementary charge, L and W are the length and width of the channel, respectively. In Eq. (1) n_0 is the background charge concentration due to random electron – hole puddles [10] and n_{BG} is the charge induced by gate bias calculate from the equation where C_{BG} is the gate capacitance per unit area taken to be 0.115 mF for 300 nm SiO_2 substrate.

We measured the electrical resistance between the source and drain as a function of the applied gate bias. As one can see in Figure 2a, the ambipolar property of graphene is preserved after irradiation within the examined irradiation dose. The observed up shift of the curves indicates increasing resistivity of graphene over a wide range of carrier concentration. The increase is especially pronounced after the compiled irradiation dose exceeds ~2500 μC/cm^2. The abrupt change of the last step may due to a difference defect mechanism from a phase change. Other factors contributing to the growth of the disorder D band can be contaminant molecules or water vapor, which dissolve under irradiation and may form bonds with the carbon atoms of the graphene lattice.

The inset to Fig. 2b shows the result of the fitting with Eqs. (1-2) of the data for SLG device before e-beam irradiation. Note that the fitting dose not cover the interval close to the charge neutrality point because this region is characterized by a large uncertainty in the data. The fitting was conducted separately for the negative and positive gate bias regions. For simplicity, we consider the fitting results from the p-type branch. The fitting gives the contact resistance of 446 Ω, the initial mobility μ=5075 cm^2/Vs, and the charge impurity concentration of $2.13\times10^{11}cm^{-2}$, which are very close to the typical values for clean graphene samples [17].

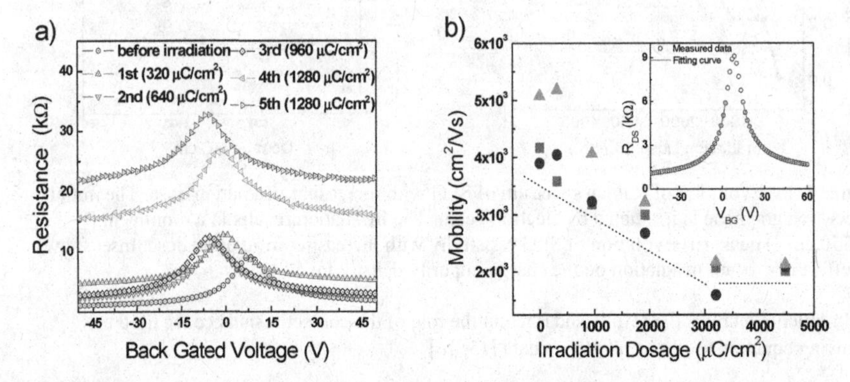

Figure 2: (a) Transfer characteristics of SLG with increasing irradiation dose. The irradiation dose for each step is indicated. (b) Mobility of three SLG devices decreases almost linearly with irradiation dosage until a certain threshold value. The inset shows fitting result for one device after the 4th irradiation step.

During the experiments the irradiated regions excluded the contacts. For this reason, the contact resistance is not expected to change. To fit our results for irradiated graphene devices we modified equation provided in Eq. (1) and (2) by adding the term R_{Ird},=(L/W)ρ_{Ird}, which is the resistance increment induced by e-beam irradiation. Fig. 2b also shows the evolution of the mobility due to e-beam irradiation for three SLG devices. We note that the mobility decreases almost linearly and drops by 50~60% over the examined irradiation dose.

It was also observed that the irradiation induced changes in the properties of SLG are reversible to some degree. The IV characteristics can be at least partially recovered by annealing or storing the devices over a long period of time in a vacuum box. The annealing may help to repair the bonds and clean the surface from the organic residues. As can be seen in Fig. 3a, the electron beam irradiation trajectory on our SLG devices is altered as the annealing is incorporated during the process. When the sample is annealed at ~510 K for a few minutes, we note the diminishing of the I(D)/I(G) ratio, as can be seen indicated in red region in Fig. 3b. To minimize such adsorbates, irradiation is conducted under vacuum conditions and then annealed. This way we are assured that the disordered D-peak appearance and removal as a result of our post annealing process is solely introduced as a result of irradiation defects or disorders created in the graphene devices. Ambient conditions alone do not lead to the disorder D-peaks. Irradiation is responsible for appearance of the disorder signatures.

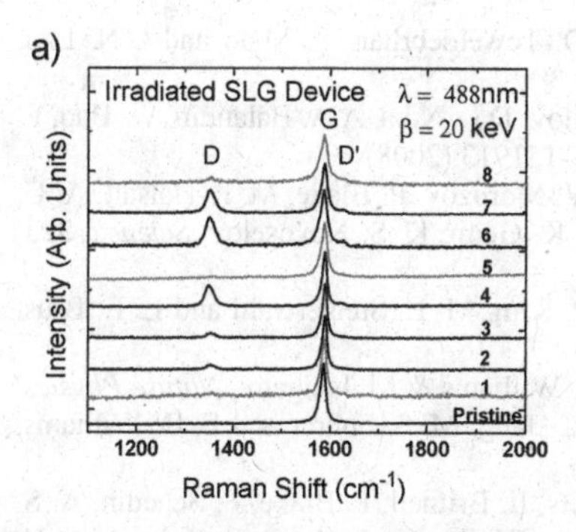

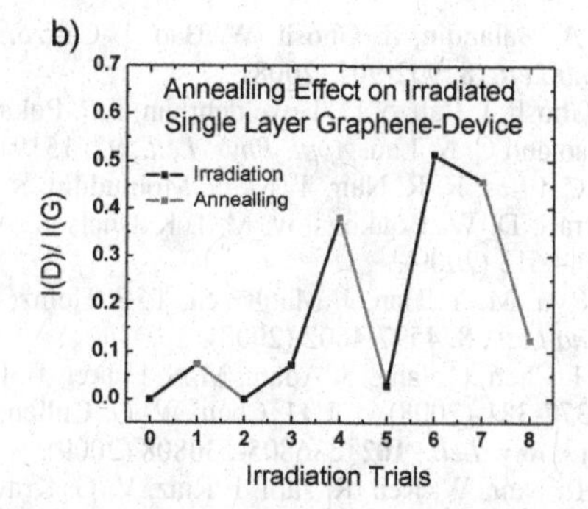

Figure 3: a) Raman spectra of SLG device showing the effects of irradiation and annealing at ~500 K. Annealing leads to disappearance of the disorder D band. b) The effect of annealing on the I(D)/I(G) ratio. The annealing steps are indicated in red between irradiation trials 1-2, 4-5, and 7-8.

CONCLUSIONS

We carried out detail investigation of the electrical and Raman spectroscopic characteristics of graphene and bilayer graphene under the electron beam irradiation. It was shown that the single layer graphene is more susceptible to e-beam irradiation than bilayer graphene. The appearance of the disorder induced D peak in graphene Raman spectrum suggests that e-beam irradiation induce defects in graphene lattice. The mobility and electrical resistivity of graphene can be varied by the e-beam irradiation over a wide range of values. The changes are reversible by annealing until some critical irradiation dose is reached. The results have important implications for fabrication of graphene nanodevices, which involve scanning electron microscopy and electron beam lithography.

ACKNOWLEDGMENTS

This work was supported, in part, by SRC – DARPA through FCRP Functional Engineered Nano Architectonics (FENA) center and DARPA Defense Microelectronics Activity (DMEA) under agreement number H94003-10-2-1003. The SEM characterization was carried out at UCR's Central Facility for Advanced Microscopy and Microanalysis (CFAMM).

REFERENCES

1. K. S. Novoselov, A. K. Geim, S. V. Morozov, D. Jiang, Y. Zhang, S. V. Dubonos, I. V. Grigorieva, A. A. Firsov, *Science*, **306**. 666 - 669 (2004).
2. Y. B. Zhang, Y. W. Tan, H. L. Stormer, and P. Kim, *Nature*, **438**, 201-204 (2005).

3. A. A. Balandin, S. Ghosh, W. Bao, I. Calizo, D. Teweldebrhan, F. Miao and C.N. Lau, *Nano Lett.*, **8**, 902-907 (2008).
4. S. Ghosh, I. Calizo, D. Teweldebrhan, E.P. Pokatilov, D.L. Nika, A.A. Balandin, W. Bao, F. Miao and C. N. Lau, *Appl. Phys. Lett.*, **92**, 151911-151913 (2008).
5. D. C. Elias, R. R. Nair, T. M. G. Mohiuddin, S. V. Morozov, P. Blake, M. P. Halsall, A. C. Ferrari, D. W. Boukhvalov, M. I. Katsnelson, A. K. Geim, K. S. Novoselov, *Science*, **323**, 610 – 613 (2009).
6. S. Ryu, M. Y. Han, J. Maultzsch, T. F. Heinz, P. Kim, M. L. Steigerwald and L. E. Brus, *Nano Lett.*, **8**, 4597-4602 (2008).
7. J. H. Chen, C. Jang, S. Adam, M. S. Fuhrer, E. D. Williams & M. Ishigami, *Nature Physics*, **4**, 377-381 (2008). ; J. H. Chen, W. G. Cullen, C. Jang, M. S. Fuhrer and E. D. Williams, *Phys. Rev. Lett.*, **102**, 236805-236808 (2009).
8. R. R. Nair, W. Ren, R. Jalil, I. Riaz, V. G. Kravets, L. Britnell, P. Blake, F. Schedin, A. S. Mayorov, S. Yuan, M. I. Katsnelson, H.-M. Cheng, W. Strupinski, L. G. Bulusheva, A. V. Okotrub, I. V. Grigorieva, A. N. Grigorenko, K. S. Novoselov, A. K. Geim, *Small*, **6**, 2877–2884 (2010).
9. D. Teweldebrhan and A. A. Balandin, *Appl. Phys. Lett.*, **94**, 013101 (2009).
10. D. Teweldebrhan and A.A. Balandin, *Appl. Phys. Lett.*, **95**, 246102 (2009).
11. G. Liu, D. Teweldebrhan, and A.A. Balandin, *IEEE Trans. Nanotechnology*, **1**, 10.1109/TNANO.2010.2087391 (2010).
12. I. Calizo, S. Ghosh, F. Miao, W. Bao, C.N. Lau and A.A. Balandin, *Solid State Communications*, **149**, 1132-1135 (2009).
13. I. Calizo, D. Teweldebrhan, W. Bao, F. Miao, C. N. Lau, and A. A. Balandin, *Journal of Physics: Conf. Ser.*, **109**, 012008 (2008).
14. A. C. Ferrari and J. Robertson, *Phys. Rev. B*, **61**, 14095-14107 (2000).
15. S. Kim, J. Nah, I. Jo, D. Shahrjerdi, L. Colombo, Z. Yao, E. Tutuc and S. K. Banerjee, *Appl. Phys. Lett.*, **94**, 062107-062109 (2009).
16. L. Liao, J. Bai, Y. Qu, Y. Lin, Y. Li, Y. Huang, and X. Duan, *PNAS*, **107**, 6711-6715 (2010).
17. S. Adam, E. H. Hwang, V. M. Galitski, and S. Das Sarma, *PNAS*, **104**, 18392-18397(2007).

Mater. Res. Soc. Symp. Proc. Vol. 1344 © 2011 Materials Research Society
DOI: 10.1557/opl.2011.1347

Self-aligned Graphene Sheets-Polyurethane Nanocomposites

Mohsen Moazzami Gudarzi[1,2], Seyed Hamed Aboutalebi[1], Nariman Yousefi[1], Qing Bin Zheng[1], Farhad Sharif[2], Jie Cao[3], Yayun Liu[3], Allison Xiao[3] and Jang-Kyo Kim[1]*
[1]Department of Mechanical Engineering, Hong Kong University of Science and Technology, Clear Water Bay, Kowloon, Hong Kong
[2]Department of Polymer Engineering, Amirkabir University of Technology, 424 Hafez Ave, Tehran, Iran
[3]Advanced Technologies, Henkel Corporation, 10 Finderne Ave. Bridgewater, NJ 08807, USA

ABSTRACT

Processing graphene and graphene polymer nanocomposites in an aqueous medium has always been a big challenge due to the hydrophobic nature of graphene (or reduced graphene oxide) nanosheets. In this work, a waterborne latex of polyurethane has been used both as the matrix material for embedding the graphene nanosheets and as a unique stabilizer to help produce an up to 5 wt% graphene/PU nanocomposites. The graphene oxide/polyurethane latex aqueous suspension is reduced *in-situ* using hydrazine, without any trace of aggregation/agglomeration upon completion of the reduction process, which would otherwise have occurred severely were PU not present. A highly aligned nanostructure is produced when graphene content is increased beyond 2 wt%, resulting in a remarkable improvement in electrical and mechanical properties of the nanocomposite. The exceptionally low electrical percolation threshold of 0.078%, as well as 21-fold and 14 fold increases in tensile modulus and strength, respectively, have been attained thanks to the alignment of graphene nanosheets in the polymeric matrix.

INTRODUCTION

Graphene, a single atomic layer of sp^2 carbon atoms has gained much attention in recent years. This unique arrangement of carbon atoms bring about exclusive electrical, thermal, magnetic and mechanical properties [1-4]. Graphene is known to be the hardest and stiffest material on earth. What adds up to these fascinating properties of graphene is the mere fact that it can be produced from cheap and naturally occurring graphite flakes [5-7]. Being one of the thinnest and strongest materials with exceptionally high electron mobility and thermal conductivity suggests that one of the most efficient ways to benefit from the properties of graphene is to incorporate them in a polymeric material to form nanocomposites [8]. Therefore, graphene oxide (GO) and reduced graphene oxide (rGO) sheets have attracted significant attention as filler for polymer nanocomposites that are now finding diverse applications [9-11].

This paper presents a simple method for the production of polymer-ultra-large graphene composites using water-borne polyurethane (PU) latex where the polymer serves both as stabilizer in the reduction step as well as the matrix material. Interparticle interactions between the ultra-large size GO sheets and PU particles are considered as the key underlying mechanism for the stabilization. The unique morphology of PU-graphene dispersion resulted in self-alignment of GO sheets with strong interfacial interactions with the matrix. The resulting composites exhibited exceptional mechanical properties and electrical conductivities with one of the lowest percolation threshold values ever reported.

EXPERIMENTAL

Ultra-large size GO (UL-GO) was synthesized based on the modified chemical method [7] using expanded graphite . The obtained GO particles were diluted using DI water (~1 mg/ml) and mildly sonicated to avoid breakage. The diameter of the as-produced GO sheets ranged between sub-micron to two hundred μm, with an average size of 32.7 ± 24.3μm [12]. The GO dispersion was mixed with aqueous emulsion of polyurethane (PU) to obtain a homogeneous aqueous dispersion. Hydrazine solution was added in the weight ratio of 3:1 to obtain reduced GO (rGO), which was then heat-treated at 80°C for 24 hr. The mixture was poured into a flat mold and dried in an oven at 50°C for 6 hr to produce composite films. The final products were highly flexible, black composite films.

Zeta potentials analyzer (Brookhaven ZetaPlus), transmission electron microscopy (TEM, JEOL 100X), scanning electron microscope (SEM, JEOL 6390F), tapping-mode atomic force microscope (AFM, Digital Instruments), four-point probe method using a resistivity/Hall measurement system (Scientific Equipment & Services) and dynamic mechanical analyzer (DMA 7, Perkin Elmer) were used to characterize the samples.

DISCUSSION

Stabilization and alignment of rGO using polyurethane dispersion

As mentioned in the experimental section, the nanocomposites were prepared by hydrazine reduction of PU/GO colloidal mixture. The reduced mixture maintained a remarkable stability, though rGO was highly hydrophobic. This phenomenon can be explained in light of the surface charge of the colloidal particles. Zeta potential is a measure of the surface electrical charges and the stability of the colloidal dispersions. The Zeta potentials of PU, GO, rGO and PU/GO colloidal mixtures have been measured and presented in Figure 1. The addition of PU to GO further increases $|\xi|$, therefore addition of PU latex particles has a pronounced effect on stabilizing GO. Figure 2 shows a photograph of GO, rGO and PU/GO dispersions one month after their initial synthesis. It can be seen that rGO was fully agglomerated but the samples containing the mixture of PU/rGO and GO alone were perfectly stable.

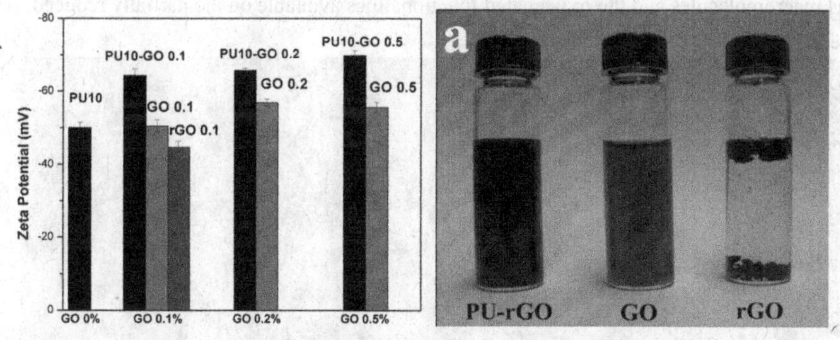

Figure 1. Zeta potentials of PU, GO, PU/GO and PU/rGO hybrid dispersions.

Figure 2. Photographs of GO, rGO and PU/rGO dispersions.

It is thought that polymer particles adhere to the graphene surface and fuse together to form a uniform polymeric coating on the GO sheet surfaces, which is mainly responsible for attributing the stability of PU/rGO colloids. Alike macromolecules, polymer particles can also be adsorbed onto the surface of GO sheets [13].This process is perhaps enthalpically favorable because of possible interaction between the hydrophilic groups on the surface of both the GO and PU particles [14].

The formation of the mentioned polymeric coating layer can also be seen in AFM and TEM images. Figures 3a and 3b depict the typical AFM scans of graphene and PU covered graphene, respectively. The height profile is included for a more quantitative treatment. It can be seen that the average step height of GO was around 0.7-1.0 nm which has been well documented as the thickness of a single layer GO sheet [5]. Interestingly, when PU is introduced to GO this height changes to 4.0-5.0 nm, as a result of adherence and fusing of latex particles onto the GO surface which lead to the formation of a uniform polymeric film.

TEM studies (Figures 3c and 3d) also support the proposed mechanism. A comparison between GO (figure 3c) and PU/rGO (figure 3d) indicated an increase in contrast when PU was introduced, which can be interpreted as an increase in the layer thickness, consistent with the AFM micrographs. The uniform nature of the formed film can be better seen in TEM pictures. Therefore, it can be said that the latex particles fuse together to form a stable and uniform film.

The freeze fracture surface of the nanocomposites have been studied using SEM. Figure 4 shows the typical micrographs for nanocomposites containing up to 5 wt% of GO sheets. The graphene sheets were seen well dispersed and no trace of aggregation/agglomeration was found in the typical fracture surfaces, given the fact that no graphene sheet was directly exposed to the fracture surface [15, 16]. This observation reflects strong interactions and adhesion between the polar PU macromolecules and the oxygenated functionalities available on the partially reduced GO sheets.

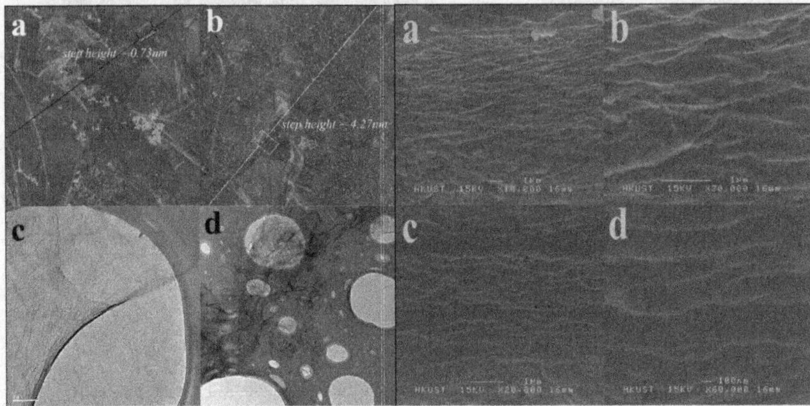

Figure 3. AFM images of a) GO and b) PU/rGO composite. TEM micrographs of c) GO and d) PU/rGO. The scale bar in c,d) is 1μm.

Figure 4. SEM images of freeze fracture surface of PU/rGO composites containing a,b) 2wt% and c,d) 5wt% graphene.

A more interesting feature associated with the fracture surfaces is the aligned nature of GO sheets, specifically in the highly-filled 5 wt% PU/rGO sample. The degree of alignment increased with increasing graphene content, therefore the UL-GO sheets tend to self-align upon evaporation of water. This is a possible scenario as a result of the very large aspect ratio of the GO sheets, approximately 30000 [12], and the low viscosity of the medium (water) both of which facilitates the mobility of nanosheets. In fact, the steric hindrance arising between the microscopically large graphene sheets of a high aspect ratio induced rearrangement into a layered structure during drying, while the low viscosity medium, such as water, and the long drying time encouraged the near-perfect alignment of graphene sheets.

Electrical conductivity

The high electrical conductivity and high aspect ratio of graphene make it an outstanding candidate for applications to avoid electrostatic and magnetic interference. The electrical conductivity versus graphene content of PU/rGO nanocomposites is presented in Figure 5. The percolation threshold of this system seems(see inset of Figure 5) to be as low as 0.078 wt%, which, to of our knowledge, is the lowest value reported for homogeneous polymer-graphene nanocomposites in the open literature. This exceptionally low value can be attributed to the highly-aligned monolayer graphene sheets with an extremely large aspect ratio as a result of the unique processing method developed in this study.

Mechanical properties of PU/rGO composites

Graphene is the strongest material ever synthesized [6], therefore, it can be used to enhance the mechanical properties of polymers. The tensile modulus and strength of neat PU and the corresponding nanocomposites are presented in Figure 6. In is noteworthy to mention that addition of only 0.3 wt% of graphene increased the tensile modulus and strength of the nanocomposites by 110% and 390%, respectively. When the graphene content was further raised to 3 wt%, remarkable 21-fold and 14-fold increases were achieved.

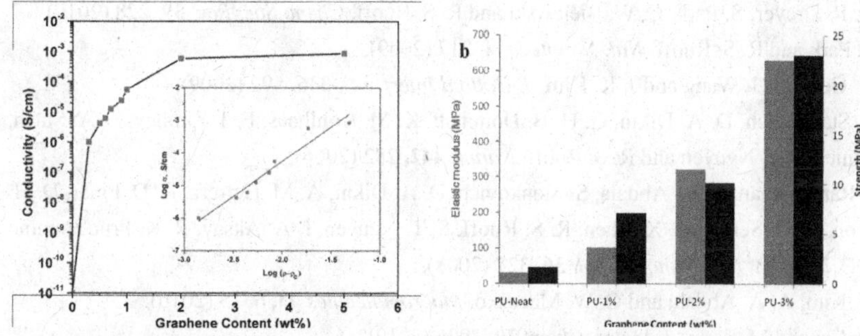

Figure 5. Electrical conductivity of PU-rGO composites as a function of graphene content, ρ.

Figure 6. Mechanical properties of PU-rGO composites as a function of graphene content.

Such remarkable improvements in mechanical properties are attributed to three interrelated factors, namely i) fine exfoliation of graphite nanoplatelets into ultra-large size, monolayer graphene sheets with high aspect ratios; ii) self-alignment of individual graphene sheets when the graphene content is above a threshold value; and iii) strong interfacial interaction between the graphene sheets and PU matrix.

CONCLUSION

Waterborne PU latex was used to produce nanocomposites containing well dispersed and highly oriented UL-GO sheets. PU not only functioned as a continuous matrix for embedding the GO sheets, but also stabilized the extremely hydrophobic rGO in the aqueous medium. It was shown that when the graphene content exceeded 2 wt%, a highly aligned layered structure was formed which mimics that of the naturally occurring nacre shells. Remarkable 21-fold and 14-fold increases in the tensile modulus and strength, respectively, were achieved by the addition of 3wt% of graphene. The composites also showed excellent electrical conductivity with an exceptionally low percolation threshold of 0.078 vol%, which is one of the lowest values reported in the literature for polymers containing graphene.

REFERENCES

[1] K. S. Novoselov, A. K. Geim, S. V. Morozov, D. Jiang, Y. Zhang, S. V. Dubonos, I. V. Grigorievaa and A. A. Firsov, *Science* **306**, 666.

[2] C. N. R. Rao, A. K. Sood, K. S. Subrahmanyam and A. Govindaraj. *Angew. Chem. Int. Ed.* **48**, 7752.

[3] A. K. Geim. *Science* **324**, 1530 (2009).

[4] C. Soldano, A. Mahmood and E. Dujardin. *Carbon* **48**, 2127 (2010).

[5] D. R. Dreyer, S. Park, C. W. Bielawski and R. S. Ruoff. *Chem. Soc. Rev.* **39**, 228 (2010).

[6] S. Park and R. S. Ruoff. *Nat. Nanotech.* **4**, 217 (2009).

[7] Y. Geng, S. J. Wang and J. K. Kim. *J. Colloid Interf. Sci.* **336,** 592 (2009).

[8] S. Stankovich, D. A. Dikin, G. H. B. Dommett, K. M. Kohlhaas, E. J. Zimney, E. A. Stach, R. D. Piner, S. T. Nguyen and R. S. Ruoff. *Nature* **442,** 282 (2006).

[9] T. Ramanathan, A. A. Abdala, S. Stankovich, D. A. Dikin, A. M. Herrera, R. D. Piner, D. H. Adamson, H. C. Schniepp, X. Chen, R. S. Ruoff, S. T. Nguyen, I. A. Aksay, R. K. Prud'Homme and L. C. Brinson. *Nat. Nanotechnol.* **36,** 327 (2008).

[10] H. Kim, A. A. Abdala and C. W. Macosko. *Macromolecules* **43,** 6515 (2010).

[11] D. Cai and M. Song. *J. Mater. Chem.* **20,** 7906 (2010).

[12] S. H. Aboutalebi, M. M. Gudarzi, Q. B. Zheng and J. K. Kim, *Adv. Funct. Mater.* in press (2011).

[13] M. M. Gudarzi and F. Sharif. *J. Colloid Inter. Sci.* **349**, 63 (2010).

[14] D. K. Chattopadhyay and K. V. S. N. Raju. *Prog. Polym. Sci.* **32**, 352 (2007).

[15] J. Li, L. Vaisman, G. Marom and J. K. Kim. *Carbon* **45**, 744 (2007).

[16] J. Li, J. K. Kim, M. L. Sham and G. Marom. *Compos. Sci. Technol.* **67**, 296 (2007)

Mater. Res. Soc. Symp. Proc. Vol. 1344 © 2011 Materials Research Society
DOI: 10.1557/opl.2011.1346

Synthesis of Graphene-CNT Hybrid Nanostructures

Maziar Ghazinejad [1,2], Shirui Guo [3], Rajat K. Paul [1], Aaron S. George [4], Miroslav Penchev [2], Mihrimah Ozkan [2], and Cengiz S. Ozkan [1]

[1] Department of Mechanical Engineering, University of California, Riverside, CA 92521, U.S.A.
[2] Department of Electrical Engineering, University of California, Riverside, CA 92521, U.S.A.
[3] Department of Chemistry, University of California, Riverside, CA 92521, U.S.A.
[4] Materials Science and Engineering Program, University of California, Riverside, CA 92521, U.S.A.

ABSTRACT

Using chemical vapor deposition technique, a novel 3D carbon nano-architecture called a pillared graphene nanostructure (PGN) is *in situ* synthesized. The fabricated novel carbon nanostructure consists of CNT pillars of variable length grown vertically from large-area graphene planes. The formation of CNTs and graphene occurs simultaneously in one CVD growth treatment. The detailed characterization of synthesized pillared graphene shows the cohesive structure and seamless contact between graphene and CNTs in the hybrid structure. The synthesized graphene-CNT hybrid has a tunable architecture and attractive material properties, as it is solely built from sp^2 hybridized carbon atoms in form of graphene and CNT. Our methodology provides a pathway for fabricating novel 3D nanostructures which are envisioned for applications in hydrogen storage, nanoelectronics, and supercapacitors.

INTRODUCTION

Graphene, a single sheet of sp^2-hybradized carbon atoms, has been attracting major attention due to its promising properties such as high charge carrier mobility, unique band structure, mechanical robustness, high thermal transport, and chemical stability [1-5]. As a result, there has been a considerable amount of theoretical and experimental research towards potential applications of graphene nanostructures in field-effect transistors, actuators, solar cells, batteries, and sensors [6-8]. Carbon nanotubes (CNT), On the other hand, have been extensively investigated over the last two decades for their exceptional electronic, optical, mechanical and chemical properties [9]. They also possess several unique features such as the ability to carry large current densities and fast electron-transfer kinetics when used as electrodes for electrochemical sensing applications. Furthermore, CNT based electrodes provide reduced reaction potential and minimum surface fouling effects and therefore offers superior performance when employed in applications for electrochemical sensing, energy storage, and photovoltaics.

It appears that for a realistic utilization of graphene and CNTs in many of the above-mentioned applications there is a need for graphene layers to have engineered architectures with sp^2-hybridized carbon atoms as building blocks. Such conceived graphene-CNT hybrid structures will combine attractive material properties of both CNTs and graphene with the capability to develop a variety of geometries. This versatility makes sp^2 hybrid carbon materials ideal candidates for a number of applications including nanoelectronics, energy conversion and storage, sensors, and displays [10-12]. The common requirements in these applications are carefully engineered architectonics, and nanoscale assembly of different building blocks of desired hybrid structure. Such material nano-engineeering is challenging due to strong Van deer Waals interaction between sp^2 graphitic materials that yields restacking of graphene layers and bundling of carbon nanotubes. Therefore, devising a fabrication methodology for spatial distribution of graphene layers and CNTs in hybrid carbon architectures is crucial and rewarding.

In this report, we describe the synthesis of graphene and carbon nanotubes into Pillared Graphene Nanostructure (PGN). Pillared graphene is a new three-dimensional carbon hybrid nanostructure comprising stacked CNT pillars on large-area graphene layers, which are envisioned for future applications in hydrogen storage and super-capacitors. The initial idea of such 3-D carbon networks was sparked by theoretical [13] and experimental [14] studies aiming at introducing novel energy storage assemblies based on tunable and large surface area nanostructures. The present study focuses on one-step CVD process for fabrication of large area pillared graphene on copper foil, via parallel growth of CNTs and graphene layers. Due to the integrated nature of the synthesis method, the resulting hybrid structure shows a very good cohesion and seamless contact between graphene and CNT building blocks. Further study of the synthesized pillared graphene facilitates application-oriented optimization of these structures.

EXPERIMENTAL DETAILS

First, SiO_2/Si substrate (area: 1 cm^2) were deposited by 600 nm thick Cu film using an E-beam evaporator (Temescal, BJD-1800). An Oxygen plasma treatment (STS Reactive Ion Etcher) was then carried out on the copper film, followed by e-beam evaporation of 3-5 nm thick Fe catalyst on the deposited copper film. Next, growth substartes were loaded into a CVD chamber and heated to 750°C under the flow of 500 sccm of Ar. All flow rates were precisely controlled by using mass flow controllers. To make the velocity profile of gas flow more uniform a total pressure of 650 Torr is maintained in the CVD tube. Once the temperature is stabilized at 750°C, first 100 sccm flow of H_2, and after 5 minutes 40 sccm flow of C_2H_2 are introduced into the tube for 10 to 20 minutes to begin the synthesis of graphene-CNT hybrid. Upon completion of the CVD growth, C_2H_2 and H_2 gas feeds were stopped, and the furnace was cooled to the room temperature under the protection of Ar gas flow. If needed, graphene-CNT hybrid films were removed from the Cu film by etching in a 1M aqueous $FeCl_3$ solution, followed by cleaning with an aqueous HCl (5%) and D.I. water solutions. Subsequently, pillared graphene films were collected from the D.I. water with a substrate of choice. Detailed characterization of the synthesized CNT-graphene hybrid films and catalyst surfaces were performed using optical microscopy, Scanning Electron Microscopy (SEM; Leo, 1550), and Transmission Electron Microscopy (TEM; Philips, CM300) with a LaB_6 cathode operated at 300kV and equipped with an X-ray energy-dispersive spectroscopy (EDS) module. TEM samples were prepared by collecting the etched graphene-CNT films suspended on D.I. water, with a TEM copper grid.

DISCUSSION

Figure 1a shows a SEM micrograph of a large-area graphene-CNT hybrid nanostructure. The SEM image indicates that the average diameter of CNTs is less than 20nm and that the CNTs have a narrow size distribution. Figure 1b shows uniform morphology of CNTs over the large area of the hybrid layer. The height of CNT pillars in this structure can be controlled by changing CVD growth parameters of growth time and flow rate of the source gas C_2H_2.

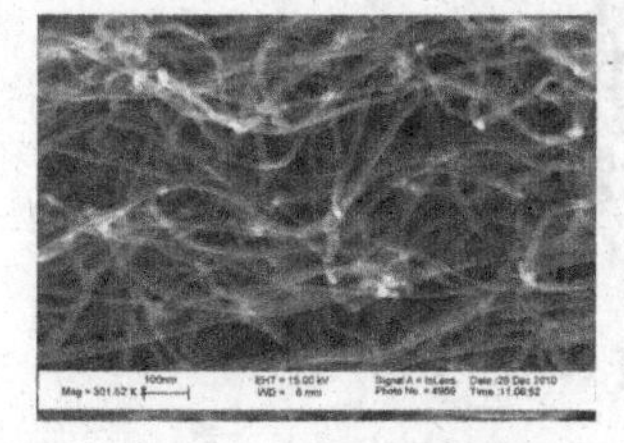

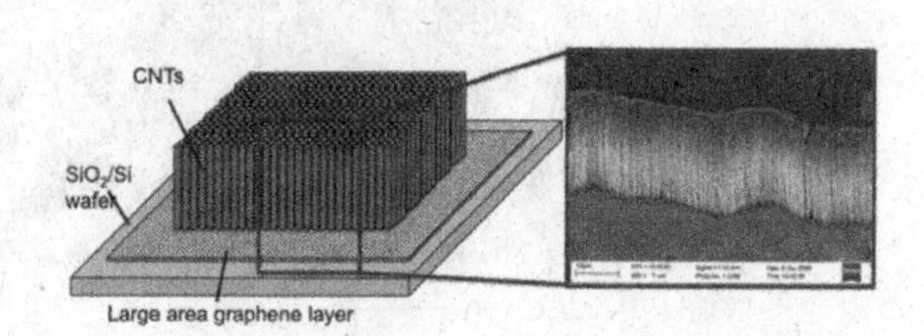

Figure 1. a) Showing a SEM image of large-scale graphene-CNT hybrid and b) A side view of the graphene-CNT hybrid showing well ordered, vertically aligned CNTs on the surface.

In order to study the graphene layers grown under the described conditions, a CVD treatment with same growth parameters was carried out on the copper foil which was only partially covered with iron on the selected areas. Upon completion of the growth the graphene sheet is etched and transferred to a SiO_2/Si substrate. Figure 2 is a typical Raman spectra taken from a graphene sheet grown on the areas without any iron deposition. The Raman spectra confirms that the growth conditions produce bi- to few layers graphene on the copper foil.

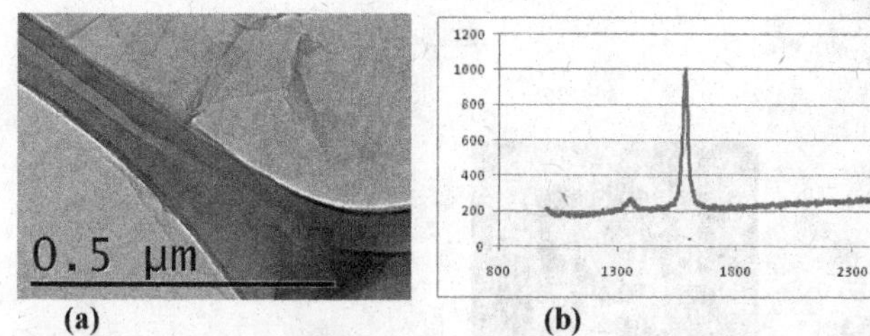

(a) (b)

Figure 2. (a) TEM image displays silk-like graphene film on carbon supported TEM copper grid; (c) Raman spectrum of the graphene film grown at the prescibed conditions shows the presence of D, G, and G' bands, confirming the presence of bi- to few layers graphene on the hybrid structure.

Figure 2a is a TEM micrograph displaying silk-like graphene film that has been grown at 750°C with acetylene, and is suspended on carbon supported TEM copper grid. Figure 2b is Raman spectrum of the graphene film grown with the procedure described in experimental section. The spectrum shows the presence of D (1340cm^{-1} to 1360cm^{-1}), G (1580cm^{-1} to 1600cm^{-1}), and G' (2680cm^{-1} to 2710cm^{-1}) bands. The Raman spectra confirms that the prescibed conditions results bi- to few layers graphene in the PGN's floor .

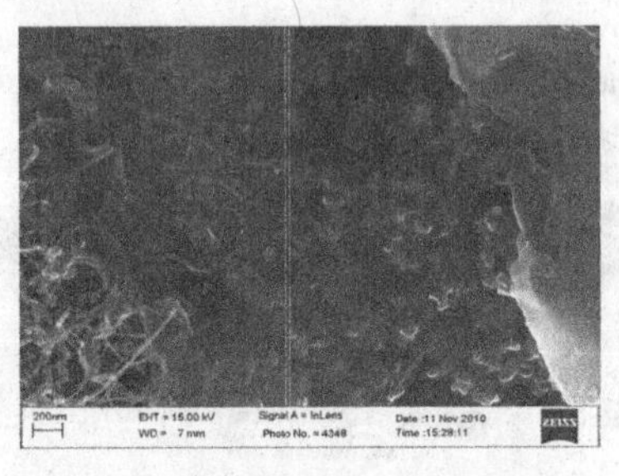

Figure 3. Bottom-veiw SEM micrograph of the backside of the peeled PGN film, showing the graphene film embedding CNT pillars roots.

Figure 3 displays a bottom-up view SEM micrograph of a large area CNT-graphene hybrid layer. To obtain this bottom-view image, the hybrid film was peeled using a double sided carbon tape and then placed onto SEM sample holder. The image displays CNT roots embedded in a few layer graphene sheet. This image indicates that the "pillars" of CNTs have good cohesion and are well connected to graphene "floor". In addition to the SEM image, we can observe that the carbon tape peeled the whole CNT-graphene hybrid rather than only CNTs, indicates a robust mechanical connection between CNTs and graphene in this graphene-CNT hybrid.

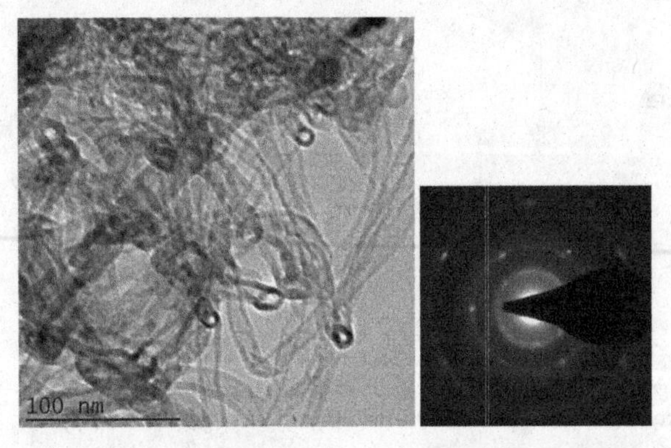

Figure 4. TEM micrograph of pillared graphene bottom-view displaying CNTs roots, at their interfaces with graphene film; and SAED diffraction pattern of pillared graphene.

Figure 4 shows a TEM micrograph of a graphene-CNT hybrid structure through the underlying graphene. This significant observation reveals that CNT pillars are directly connected to the graphene film. The hollow circular patterns at the interface of the CNTs and graphene point toward the probability of the direct growth of CNTs from graphene layers. This indicates a smooth transition between two different carbon allotropes in the hybrid structure. The typical six-fold symmetry in diffraction spots supports the presence of mono- to few-layer underlaying graphene film. Moreover, the concentric circles in the selected area diffraction pattern of pillared graphene sample are originated by overlapping atomic carbon planes, in form of CNT and graphene, with random orientation relative to the incident electron beam. The depicted seamless contact between graphene and carbon nanotubes in the hybrid structure is beneficial for the use of this 3-D carbon structure in several applications including interconnects, electrodes for supercapacitors, and field-emitting device technology.

CONCLUSIONS

In the present study, we report the synthesis of a carbon hybrid structure, via CVD growth of graphene layers and carbon nanotubes. The synchronous growth methodology is compatible with industrial thin-film fabrication technologies, since it does not involve any unconventional laboratory steps. The graphene-CNT hybrid is a remarkable carbon nanostructure with tunable nano-architectures, making it a suitable application-oriented design of hierarchical graphene structures. Characterization of synthesized pillared graphene shows the cohesive structure with seamless contact between graphene and CNTs in the hybrid. Through its tunable spatial geometry and attractive material properties, this carbon hybrid has very promising potential for integration into nanoelectronics, energy storage, and even electronic cooling and thermal management.

REFERENCES

1. K.S. Novoselov, A.K. Geim, S.V. Morozov, D. Jiang, M.I. Katsnelson, I.V. Grigorieva, S.V. Dubonos, A.A. Firsov, Nature, 438 (2005) 197-200.
2. K.S. Novoselov, A.K. Geim, S.V. Morozov, D. Jiang, Y. Zhang, S.V. Dubonos, I.V. Grigorieva, A.A. Firsov, Science, 306 (2004) 666-669.
3. Y. Zhang, Y.-W. Tan, H.L. Stormer, P. Kim, Nature, 438 (2005) 201-204.
4. F. Schedin, A.K. Geim, S.V. Morozov, E.W. Hill, P. Blake, M.I. Katsnelson, K.S. Novoselov, Nature Materials, 6 (2007) 652-655.
5. J.-H. Chen, C. Jang, S. Xiao, M. Ishigami, M.S. Fuhrer, Nat Nano, 3 (2008) 206-209.
6. X. Wang, L. Zhi, K. Mullen, Nano Letters, 8 (2007) 323-327.
7. J. Lin, D. Teweldebrhan, K. Ashraf, G. Liu, X. Jing, Z. Yan, R. Li, M. Ozkan, R.K. Lake, A.A. Balandin, C.S. Ozkan, Small, 6 (2010) 1150-1155.
8. J. Yan, T. Wei, B. Shao, Z. Fan, W. Qian, M. Zhang, F. Wei, Carbon, 48 (2009) 487-493.
9. S. Iijima, Nature, 354 (1991) 56-58.
10. D.H. Lee, J.E. Kim, T.H. Han, J.W. Hwang, S. Jeon, S.-Y. Choi, S.H. Hong, W.J. Lee, R.S. Ruoff, S.O. Kim, Advanced Materials, 22 (2010) 1247-1252.
11. H.Y. Jeong, D.-S. Lee, H.K. Choi, D.H. Lee, J.-E. Kim, J.Y. Lee, W.J. Lee, S.O. Kim, S.-Y. Choi, Applied Physics Letters, 96 (2010) 213105-213103.
12. D. Yu, L. Dai, The Journal of Physical Chemistry Letters, 1 (2009) 467-470.
13. G.K. Dimitrakakis, E. Tylianakis, G.E. Froudakis, Nano Letters, 8 (2008) 3166-3170.
14. R.K. Paul, M. Ghazinejad, M. Penchev, J. Lin, M. Ozkan, C.S. Ozkan, Small, 6 (2010) 2309-2313.

Mater. Res. Soc. Symp. Proc. Vol. 1344 © 2011 Materials Research Society
DOI: 10.1557/opl.2011.1365

Large-Area Industrial-Scale Identification and Quality Control of Graphene

Craig M. Nolen[1], Giovanni Denina[2], Desalegne Teweldebrhan[1], Bir Bhanu[2], and Alexander A. Balandin[1]
[1]Nano-Device Laboratory, Department of Electrical Engineering and Materials Science and Engineering Program, Bourns College of Engineering, University of California – Riverside, Riverside, California 92521, USA
[2]Visualization and Intelligent Systems Laboratory, Department of Electrical Engineering, Bourns College of Engineering, University of California – Riverside, Riverside, California 92521, USA

ABSTRACT

A large-area graphene layer identification technique was developed for research and industrial applications. It is based on the analysis of optical microscopy images using computational image processing algorithms. The initial calibration is performed with the micro-Raman spectroscopy. The method can be applied to the wafer-scale graphene samples. The technique has the potential to be the gateway in the development of fully automated statistical process control methods for the next generation thin-film materials used by the semiconductor industry. The proposed technique can be applied to graphene on arbitrary substrates and used for other atomically thin materials.

INTRODUCTION

Discovery of graphene, an atomic monolayer of carbon [1] stimulated major interest within the scientific and engineering community owing to material's exceptional intrinsic electronic [1] and thermal properties [2]. Research evolved quickly in the graphene field, with recent advancements in large-area growth methods of up to 30 inch rolls from chemical vapor deposition (CVD) on Cu [3] and transfer of grown graphene onto dielectric substrates [4]. Practical applications of graphene require a method for large-area identification of graphene and its quality control. CVD grown graphene wafers can have regions with different number of atomic planes and crystalline orientation of the grains [5]. The properties of graphene and few-layer graphene samples depend strongly on the number of atomic layers. For example, thermal conductivity [2] and electron mobility differ significantly between graphene [6-8] and few-layer graphene [9, 10]. The large-area consistency of layering is critical for many graphene-based electronic devices due to their sensitivity to fluctuations in the intrinsic properties [11]. Optical transparency of single and few layer graphene complicate the large-scale recognition and quality control [12]. This is especially true when transferring grown or exfoliated graphene onto various dielectric substrates [13]. Available techniques are not suitable for the large-area graphene recognition [12, 14-20].

Here we present a method for graphene layer recognition with high accuracy, high throughput, complete automation, and a way for providing statistical analysis for quality control at the industrial level for large-scale integration [21, 22]. This technique is cheap, robust, time-effective, and efficient at achieving automated high throughput large-area graphene layer recognition. Our metrology tool consists of modifying the typical optical contrast method of counting graphene layers for recognition by analyzing captured images with image processing algorithms. The image processing used for graphene layer characterization was implemented with MATLAB codes. Raman spectroscopy is used as a calibration tool for the method. The proposed technique can be extended to various substrates, surfaces, interfaces and atomically thin films produced from various materials, including topological insulators.

EXPERIMENT AND RESULTS

The experimental setup for this technique is laid out in the following number of steps. First, a graphene sample with few layered graphene regions is produced, by either mechanical exfoliation from highly oriented pyrolythic graphite (HOPG) placed on top of a 300-nm SiO_2/Si substrate or grown by CVD [12]. Second, two optical microscopy images are captured; one of these few layered graphene regions and the other of only the substrate which are obtained through a camera mounted to an optical microscope. Prior to processing this image, a calibration procedure is undertaken where in this case Raman spectroscopy is used to verify the number of atomic planes and matched with correspondence for each layer of graphene that is intended to be detected or identified [14]. With this information, image processing of the image is completed to detect and recognize the regions of each individual graphene layered regions.

Through computational processing, an image is parsed into individual pixels each assigned with a red, green, and blue (RGB) value in the range of 0 to 255. Next, non-uniform light caused from the confocal optical microscopy lens is extracted from the substrate only image and subtracted from the image with few layered graphene regions to increase the accuracy and detection of graphene layered regions to surpass short range limitations [21]. Then, the background is subtracted including; substrate, bulk graphitic layers, and polymers / residues which are all removed leaving only few layered graphene regions of interest [21]. Next, each pixel in the entire image is converted from three RGB values to one scalar grayscale value per pixel to provide a more simplistic approach for classification of each individual graphene layer contrast ranges (Equation 1.1).

$$I_n(x,y)=\begin{cases}1 & L1_{\min}\le \Delta I_1(x,y)\le L1_{\max}\\ 2 & L2_{\min}\le \Delta I_2(x,y)\le L2_{\max}\\ 3 & L3_{\min}\le \Delta I_3(x,y)\le L3_{\max}\\ 4 & L4_{\min}\le \Delta I_4(x,y)\le L4_{\max}\\ 0 & other\_values\end{cases} \quad (1.1)$$

Next, masks are produced where regions are classified, assigned a unique pseudo color (Equation 1.2), and labeled with a designated number of atomic planes extracted previously from initial calibration which in this case was Raman spectroscopy.

$$I_n(x,y)=\begin{cases} 1 & \Delta I_1(x,y)= \text{Red} \\ 2 & \Delta I_2(x,y)= \text{Green} \\ 3 & \Delta I_3(x,y)= \text{Blue} \\ 4 & \Delta I_4(x,y)= \text{Yellow} \\ 0 & other\_values \end{cases} \quad (1.2)$$

Finally, post processing of each mask consisting of classified graphene layered regions are passed through a median filter to remove impulse noise and to remove grainy portions of inaccurately identified graphene layers where this filter effectively puts a limitation for the size of minimum number of pixels that must be clustered together in order to be not be filtered out [21]. Then each mask is placed back over the original captured optical microscopy image, where the completed result is shown in Figure 1 in the left panel with its corresponding Raman spectrum.

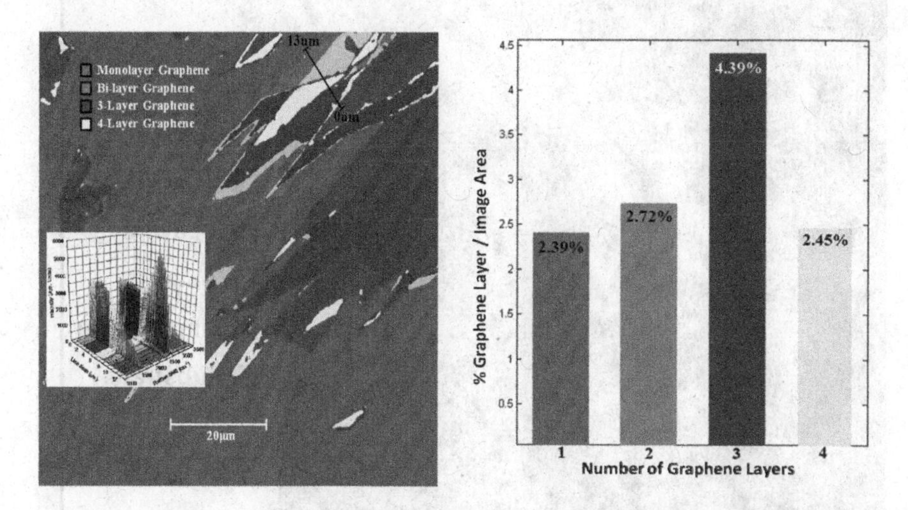

Figure 1: Processed optical microscopy image with identified graphene layered regions derived from exfoliation of bulk HOPG placed on top of a SiO_2/Si substrate (left panel). Percentage of each graphene layer throughout the entire image is displayed on a bar chart (right panel).

Since the captured image of the few layered graphene regions has been analyzed pixel by pixel a wide variety of analytical statistical interpretation can be performed including percentage of each graphene layered region over the entire image area as shown in Figure 1 in the right panel. Furthermore, this processed image can be manipulated in a number of different ways including a unique representation of a 2-Dimensional perspective of a 3-Dimensional optical microscopy image processed where the z-axis is contrast intensity and projected light is illuminated across the surface shown in Figure 2. When looking closely at this figure one can also get more information of this image set in a different representation where the light strewn across the surface shows acute differences in the surface topology showing previously minuscule differences as a much more obvious change potentially leading to the surface roughness depending on the resolution of the image.

The optical image analysis part of our experiment used image processing algorithms by using MATLAB with an extended toolbox to detect and characterize this material where the procedure is outlined in Nolen et al. [21]. This procedure used customized image processing algorithms including heuristic thresholding, background subtraction, non-uniform light subtraction, and a median filter [23, 24]. Ultimately we were able to extract out individual graphene layers for a multitude of uses e.g., percent detected over the entire image, potential for size and shape detection and further statistical interpretation and quick automated analysis.

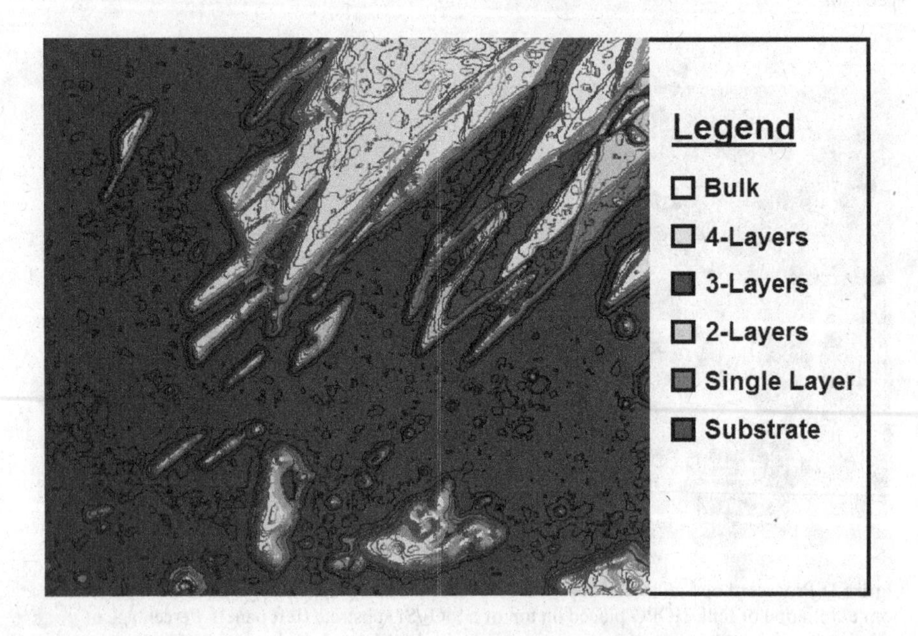

Figure 2: Processed optical microscopy image classifying both graphene layered regions as well as the bulk and substrate. The z-axis is contrast intensity where projected light is illuminated across the surface.

CONCLUSIONS

We have developed an automated large-area graphene layer identification and quality control technique suitable for large graphene wafers (inch size). The technique can be expanded for use with graphene on different substrates [25, 26], different forms of graphene i.e., exfoliated, deposited, or grown graphene, as well as for the use with other thin-film quasi-2D material systems such as atomically -thin topological insulators, i.e., Bi_2Te_3, Bi_2Se_3, Sb_2Te_3 [27, 28]. The technique can be locally calibrated with Raman spectroscopy or atomic force microscopy. Incorporation of this automated graphene layer detection technique and development of other automated quality control techniques will help to improve the quality of graphene wafers.

ACKNOWLEDGMENTS

The work in Nano-Device Laboratory (NDL) was supported, in part, by Defense Advanced Research Projects Agency (DARPA) – Semiconductor Research Corporation (SRC) Center on Functional Engineered Nano Architectonics (FENA).

REFERENCES

1. A. K. Geim and K. S. Novoselov, Nature Materials, **6**, 183-191 (2007).
2. A.A. Balandin, Nature Materials, **10**, 569 - 581 (2011).
3. S. Bae, H. Kim, Y. Lee, X. Xu, J. Park, Y. Zheng, J. Balakrishnan, T. Lei, H. R. Kim and Y. I. Song, Nat. Nano., **5**, 574-578 (2010).
4. A. Reina, J. Xiaoting, J. Ho, D. Nezich, H. Son, V. Bulovic, M. S. Dresselhaus, and J. Kong, Nano Lett., **9**, 30-35 (2008).
5. X. Li, C. W. Magnuson, A. Venugopal, J. An, J. W. Suk, B. Han, M. Borysiak, W. Cai, A. Velamakanni, Y. Zhu, L. Fu, E. M. Vogel, E. Voelkl, L. Colombo, and R. S. Ruoff, Nano Lett., **10**, 4328-4334 (2010).
6. S. Morozov, K. S. Novoselov, A. K. Geim, *et al.*, Phys. Rev Lett., **100**, 016602. (2008)
7. A. A. Balandin, S. Ghosh, I. Calizo *et al.*, Nano Lett., **8**, 3, 902-907. (2008)
8. D. L. Nika *et al.*, Phys. Rev. B, **79**, 155413. (2009)
9. K. S. Novoselov *et al.*, Science, **306**, 666-669. (2004)
10. S. Ghosh, W. Bao, D. L. Nika, S. Subrina, E. P. Pokatilov, C. N. Lau, A. A. Balandin, Nat. Materials,**9**, 555-558 (2010).
11. F. Schwierz, Nat. Nano., **5**, 487-496 (2010).
12. P. Blake, E. W. Hill, A. H. C. Neto, K. S. Novoselov, D. Jiang, R. Yang, T. J. Booth, and A. K. Geim, Appl. Phys. Lett., **91**, 063124 (2007).
13. Y. Lee, S. Bae, H. Jang, S. Jang, S. Zhu, S. H. Sim, Y. I. Song, B. H. Hong, and J. Ahn, Nano Lett., **10**, 490-493 (2010).
14. A. C. Ferrari, J. C. Meyer, V. Scardaci, C. Casiraghi, M. Lazzeri, F. Mauri, S. Piscanec, D. Jiang, K. S. Novoselov, S. Roth, and A. K. Geim, Phys. Rev. Lett., **97**, 187401 (2006).

15. A. Gupta, G. Chen, P. Joshi, S. Tadigadapa, and P.C. Eklund, *Nano Lett.*, **6**, 2667-2673 (2006).
16. C. Berger, Z. Song, T. Li, X. Li, A. Y. Ogbazghi, R. Feng, Z. Dai, A. N. Marchenkov, E. H. Conrad, P. N. First, and W. A. de Heer, J. *Phys. Chem. B*, **108**, 19912–19916 (2004).
17. C. Virojanadara, M. Syvajarvi, R. Yakimova, L. I. Johansson, A. A. Zakharov, and T. Balasubramanian , *Phys. Rev. B*, **78**, 245403 (2008).
18. E. Stolyarova, K. T. Rim, S. Ryu, J. Maultzsch, P. Kim, L. E. Brus, T. F. Heinz, M. S. Hybertsen, and G. W. Flynn, *PNAS*, **104**, 9209-9212 (2007).
19. H. Hibino, H. Kageshima, F. Maeda, M. Nagase, Y. Kobayashi, and H. Yamaguchi, *Phys. Rev. B*, **77**, 075413 (2008).
20. P.E. Gaskell, H. S. Skulason, C. Rodenchuk, and T. Szkopek, *Appl. Phys. Lett.*, **94**, 143101 (2009).
21. C. M. Nolen, G. Denina, D. Teweldebrhan, B. Bhanu, and A. A. Balandin, *ACS Nano*,**5**, 914-922 (2011).
22. C. M. Nolen, D. Teweldebrhan, G. Denina, B. Bhanu, and A. A. Balandin, *ECS Trans.*, **33**, 201 (2010).
23. R. C. Gonzalez and R. Woods, Digital Image Processing 3rd ed., Pearson Edu. Inc. (2008)
24. E. Levy, D. Peles, M. Opher-Lipson, and S. G. Lipson, *Applied Optics*, **38**, 4, 679-683 (1999).
25. G. Teo, H. Wang, Y. Wu, Z. Guo, J. Zhang, Z. Ni, and Z. Shen, *J. Appl. Phys.*,**103**, 124302 (2008).
26. I. Calizo, S. Ghosh, W. Bao, F. Miao, C. N. Lao, and A. A. Balandin, *Solid State Commun.*, **149**, 1132-1135 (2009).
27. D. Teweldebrhan, V. Goyal, and A. A. Balandin, *Nano Lett.*, **10**, 1209 (2010).
28. D. Kong, W. Dang, J. J. Cha, H. Li, S. Meister, H. Peng, Z. Liu, Y. Cui, *Nano Lett.*, **10**, 2245-2250 (2010).

Mater. Res. Soc. Symp. Proc. Vol. 1344 © 2011 Materials Research Society
DOI: 10.1557/opl.2011.1355

Rapid large-scale Characterization of CVD Graphene Layers on Glass using Fluorescence Quenching Microscopy

Jennifer Reiber Kyle[1], Ali Guvenc[1], Wei Wang[2], Jian Lin[3], Maziar Ghazinejad[3], Cengiz Ozkan[2,3], and Mihrimah Ozkan[1]

[1]Department of Electrical Engineering, University of California-Riverside, Riverside, CA 92521, USA.

[2]Department of Materials Science & Engineering, University of California-Riverside, Riverside, CA 92521, USA.

[3]Department of Mechanical Engineering, University of California-Riverside, Riverside, CA 92521, USA.

ABSTRACT

The exceptional electrical, optical, and mechanical properties of graphene make it a promising material for many industrial applications such as solar cells, semiconductor devices, and thermal heat sinks. However, the greatest obstacle in the use of graphene in industry is high-throughput scaling of its production and characterization. Chemical-vapor deposition growth of graphene has allowed for industrial-scale graphene production. In this work we introduce complimentary high-throughput metrology technique for characterization of chemical-vapor deposition-grown graphene. This metrology technique provides quick identification of thickness and uniformity of entire large-area chemical-vapor deposition-grown graphene sheets on a glass substrate and allows for easy identification of folds and cracks in the graphene samples. This metrology technique utilizes fluorescence quenching microscopy, which is based on resonant energy transfer between a dye molecule and graphene, to increase allow graphene visualization on the glass substrate and increase the contrast between graphene layers.

INTRODUCTION

Graphene is promising for many industrial applications due to its exceptional electrical, optical, thermal, and mechanical properties.[1,2] However, its use in industry is still limited because it is difficult to produce and characterize on a large scale. Graphene produced by mechanical exfoliation is of excellent quality but extremely low yield. To overcome this limitation, a new method of graphene production was developed. In this technique, graphene is grown via chemical vapor deposition (CVD) of carbon atoms on metallic substrates. CVD-grown graphene can be produced as very large sheets, with the size of the sample limited only by the substrate and growing tube.

The same characteristics that make graphene extraordinary also make it difficult to characterize. A single-layer graphene sample is only ~0.4nm thick[3] and absorbs only 2.3% of incident light.[4] Therefore, on a standard substrate such as a glass microscope slide, graphene cannot be adequately visualized in a microscope using reflection or transmission optical microscopy. Common methods for characterizing graphene thickness are Raman microscopy,[5] atomic force microscopy[6] and transmission electron microscopy. While these techniques offer insight into the atomic-scale quality of graphene samples, they are slow and limited to characterizing small regions. To overcome these issues, large-scale optical graphene metrology techniques have been developed that identify the layers of graphene immobilized oxidized silicon (Si/SiO_2) substrates based on their color contrast.[7,8] Although Si/SiO_2 substrates offer a simple method for improving the visibility of graphene, they complicate the development of a metrology technique suitably robust for industrial use. This is due to the fact that the color sensitivity of cameras changes between camera models and depends on the illumination intensity. Therefore, the color contrast used to identify graphene layers changes between microscopes and slowly changes on the same microscope as the illumination intensity varies. Therefore, metrology techniques that rely on Si/SiO_2 must be calibrated often, a step that requires Raman spectroscopy to identify each individual graphene layer. Finally, many industrial applications, including solar cells and electronic systems on PCB boards, require metrology measurements of graphene samples on substrates other than Si/SiO_2.

In this research project we measure the thickness and uniformity of large chemical-vapor deposition-grown graphene samples on microscope glass slides using fluorescence quenching microscopy (FQM). FQM is a novel technique for visualizing graphene that is based on Förster resonant energy transfer from a fluorescent dye to graphene.[9] FQM offers improvements over conventional graphene imaging techniques because it can be performed on arbitrary substrates, imaging time is short, large areas can be measured, and the imaging equipment (a fluorescent microscope) is widely available. Fluorescence quenching is visualized by spin-coating a solution of polymer mixed with fluorescent dye onto the graphene then viewing the sample under a fluorescence microscope. Because graphene quenches the dye while the substrate does not, graphene regions are identified by dark regions in the fluorescence image. Currently, FQM has only been used to visualize graphene-oxide and exfoliated graphene samples and has been unable to achieve quantitative characterization of the graphene samples, specifically, counting graphene layers.

Using the results from our fluorescence quenching calibration, we quantify the thickness and uniformity of an entire CVD-grown sample. This is achieved by creating a large-scale, high-resolution fluorescence montage image of the entire graphene sample using a microscope with a motorized stage. We segment the image based on graphene layer thickness using histogram-based segmentation and represent each graphene layer with a unique color. This method allows quick and easy identification of graphene layers in a large area on arbitrary substrates.

THEORY

CVD graphene samples were grown on pretreated copper foil in a quartz-tube furnace chamber. The copper was annealed at 1000°C for 30 minutes, then methane was introduced to the chamber under 20 Torr for 20 minutes, and then the chamber was cooled to 25°C at a 20°C/minute rate. A layer of PMMA was introduced on the graphene samples by drop-coating and then heated at 120°C for 10 minutes. The copper foil was then etched in iron(III) chloride ($FeCl_3$) aqueous solution (0.5M) and rinsed thoroughly with hydrochloric acid (3%) and DI water, respectively. Floating CVD graphene was then picked up with a cleaned microscope glass slide.

The dye mixture was prepared by adding 0.01wt% 4-(Dicyanomethylene)-2-methyl-6-(4-dimethylaminostyryl)-4H-pyran (DCM, Sigma Aldrich) to 10mL 1.0wt% PMMA (Mw ~120,000) dissolved in toluene (>99.5%, Fisher Chemical). To coat the CVD graphene with the dye layer, this mixture was passed through a 0.22μm filter and flooded onto the substrate, then the substrate was spun at 3000rpm for 60 seconds with a 2-second ramp. Measurements with a Veeco Dektak 8 surface profilometer showed the dye layer thickness to be ~30nm.

Fluorescence imaging was performed using a BD Pathway 855 HT confocal microscope equipped with an arc-lamp light source and a CCD camera. The filters used were a 470nm (+/-40nm) bandpass excitation filter, 520nm dichroic mirror, and a 542nm (+/-27nm) bandpass emission filter. BD AttoVision software, which is provided with the Pathway microscope, was used to control the mechanical stage and collect montage images.

DISCUSSION

Montage Creation and Correction

The fluorescence montage image of single-layer CVD graphene is shown in Figure 1. Because the illumination across one image is not completely uniform, individual images in the montage image can be identified by their dark outlines. For accurate measurement of intensity values used in the segmentation step, this effect must be removed.

Figure 1. Original FQM montage image.

Non-uniform illumination can be corrected using the standard microscopy flatfield correction technique. Each area in the montage image that represents an individual image is corrected using

$$I_{flat}(x, y) = \frac{I_{original}(x, y)}{I_{correction}(x, y)} \times \overline{I_{correction}} \quad (1)$$

,

where $I_{correction}$ is a correction image is created by imaging a uniform fluorescence sample such as a dye layer covering a bare substrate. In the corrected montage image (Figure 2), the non-uniform illumination has been entirely corrected.

Figure 2. FQM montage image after correction for uneven illumination.

Graphene Layer Identification

Identification of the graphene layers is achieved by histogram-based segmentation based on contrast relative to the substrate. The first step in identifying graphene layers is measuring the background intensity. This is achieved by analyzing the image histogram. Two major peaks are apparent in the histogram of the corrected fluorescence image of the CVD graphene sample (Figure 3). The peak at higher fluorescence intensities represents the substrate while the peaks at lower intensities represent the graphene. The background intensity, $I_{background}$, is the intensity value that correlates to the apex of the background peak in the histogram.

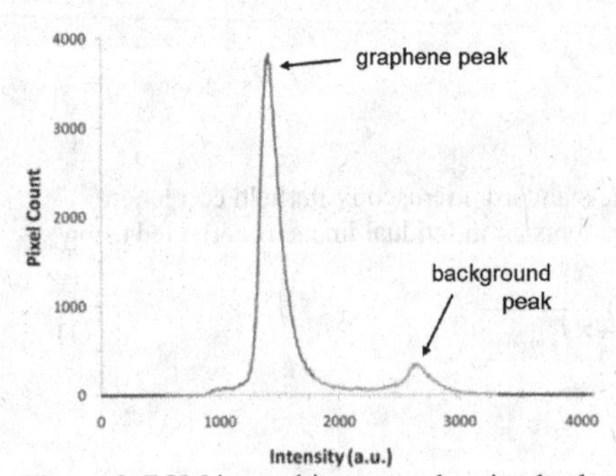

Figure 3. FQM image histogram showing background and graphene peaks.

Once the background intensity is determined, the contrast value for each pixel is calculated according to

$$C(x,y)=\frac{I_{background}-I(x,y)}{I_{background}}. \quad (2)$$

Next, the image is segmented according to the pixel contrast value. Our measurements on multiple graphene samples found that for a 30nm-thick dye layer, the contrast range for single-layer graphene is 0.35-0.58, the contrast range for two-layer graphene is 0.58-0.75, and the contrast range for 3 or more graphene layers is 0.75-0.8. Applying this segmentation algorithm to the corrected montage fluorescence image produces the segmented image shown in Figure 4. In this image, the graphene layers are portrayed using unique colors. The segmented shows that the graphene sample is entirely single-layer graphene with some easily identifiable cracks and folds with two-layer graphene. Raman microscopy measurements on the graphene sample and folds verified that the sample is primarily single-layer graphene with two-layer graphene at the folds.

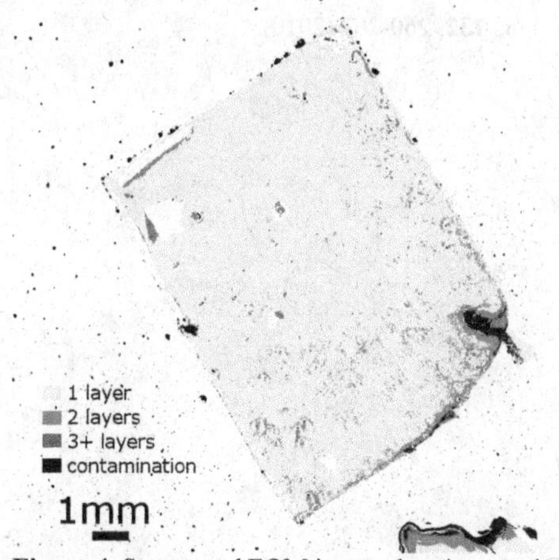

Figure 4. Segmented FQM image showing graphene layers as well as contamination.

CONCLUSIONS

In this work we have developed a large-scale metrology method for measuring the thickness and uniformity of entire CVD graphene samples. This method utilizes FQM to increase the contrast between the graphene layers and the substrate and histogram-based segmentation to identify the graphene layers. The large-scale metrology technique described in this work allows for fast and accurate evaluation of the quality of CVD graphene samples. The repeatability and flexibility of this technique make it promising for many industrial applications.

REFERENCES

1. A. Castro Neto, F. Guinea, N. Peres, K. Novoselov, A. Geim, Rev. Mod. Phys. **81**, 109-162 (2009).
2. A. Geim, K. Novoselov, Nat. Mater. **6**, 183-191 (2007).
3. K. Novoselov, A. Geim, S. Morozov, D. Jiang, Y. Zhang, S. Dubonos, I. Grigorieva, A. Firsov, Science **306**, 666-669 (2004).
4. R. R. Nair, P. Blake, A. N. Grigorenko, K. S. Novoselov, T. J. Booth, T. Stauber, N. M. R. Peres, A. K. Geim, Science **320**, 1308 (2008).
5. A. C. Ferrari, J. C. Meyer, V. Scardaci, C. Casiraghi, M. Lazzeri, F. Mauri, S. Piscanec, D. Jiang, K. S. Novoselov, S. Roth, A. K. Geim, Phys. Rev. Lett. **97**, 187401 (2006).
6. S. Stankovich, D. A. Dikin, G. H. B. Dommett, K. M. Kohlhaas, E. J. Zimney, E. A. Stach, R. D. Piner, S. T. Nguyen, R. S. Ruoff, Nature **442**, 282-286 (2006).
7. A. Reina, S. Thiele, X. Jia, S. Bhaviripudi, M. Dresselhaus, J. Schaefer, J. Kong, Nano. Res. **2**, 509-516 (2009).
8. C. M. Nolen, G. Denina, D. Teweldebrhan, B. Bhanu, A. A. Balandin, ACS Nano **5** (2) 914-922 (2011).
9. J. Kim, L. Cote, F. Kim, J. Huang, J. Am. Chem. Soc. **132**, 260-267 (2010).

Graphene and Graphene Composites: Properties and Applications

Mater. Res. Soc. Symp. Proc. Vol. 1344 © 2011 Materials Research Society
DOI: 10.1557/opl.2011.1369

Thermal Properties of Graphene and Carbon Based Materials: Prospects of Thermal Management Applications

Suchismita Ghosh[1] and Alexander A. Balandin[2]
[1]Intel Corporation, Hillsboro OR 97124, U.S.A.
[2]Nano-Device Laboratory, Electrical Engineering Department and Materials science and Engineering Program, University of California Riverside, Riverside, CA 92521, U.S.A.

ABSTRACT

In recent years, there has been an increasing interest in thermal properties of materials. This arises mostly from the practical needs of heat removal and thermal management, which have now become critical issues for the continuing progress in electronic and optoelectronic industries. Another motivation for the study of thermal properties at nanoscale is from a fundamental science perspective. Thermal conductivity of different allotropes of carbon materials span a uniquely large range of values with the highest in graphene and carbon nanotube and the lowest in amorphous or disordered carbon. Here we describe the thermal properties of graphene and carbon-based materials and analyze the prospects of applications of carbon materials in thermal management.

INTRODUCTION

As the electronic device feature size approaches a few-nanometer length scale, the increased power densities and high chip temperature hinders reliable performance of integrated circuits [1-2, 3]. While this is an issue at the chip level and posing a problem for the circuit designers, device designers have started facing problems within individual transistors and this gives rise to thermal management issues. For a wide range of devices such as complementary metal-oxide silicon CMOS and high electron mobility transistors, excessive heating severely impedes the operations. One of the main reasons behind this is the nanoscale device feature size approaching the phonon mean free path (MFP). At such a length scale, phonon boundary scattering starts dominating the three phonon Umklapp scattering. Acoustic phonons having large group velocities are the ones which contribute mostly to thermal conductivity as opposed to optical phonons with smaller group velocity. When conventional design is power constrained, in order to maintain optimum device performance, one has to take into account engineering of material parameters or structural geometry so that heat can be removed efficiently. One possible solution to the thermal issues is to find a material with very high thermal conductivity so that it can be integrated with Si based complementary metal-oxide-semiconductor (CMOS) technology and three-dimensional (3D) electronics for efficient heat removal [4].

There has been a growing interest in the thermal transport of individual nanostructures as well as nanostructure-based devices. A material's heat conduction ability is ingrained in its atomic structure and study of thermal properties of nanostructures can elucidate basic material-characteristics. Thermal transport at nanoscale is significantly different from that as macroscale [5]. Owing to increased phonon boundary scattering and changes in phonon dispersion, nanostructures like nanowires do not conduct heat as efficiently as crystalline bulk materials [6,

7]. In two-dimensional (2D) and one-dimensional (1D) crystals, the intrinsic thermal conductivity, K can reach infinitely large values as has been shown theoretically [8, 9]. In bulk materials, the crystal anharmonicity can restore thermal equilibrium. But, interestingly in 2D materials, the introduction of disorder or a limitation on the size or width of the sample can bind the values of K within *finite* limits. These works have motivated investigation of thermal transport in materials with dimensions in the order of nanoscale or atomic scale [10, 11].

Graphene, a single atomic layer of carbon was isolated in its free state through mechanical cleavage of bulk graphite only a few years ago [12] and exhibits among other interesting properties, a very high room temperature carrier mobility in the order of 15000 $cm^2V^{-1}s^{-1}$ [13, 14]. The exotic properties of graphene and its identification through Raman spectroscopy [15, 16] motivated the first thermal conductivity measurements of these 2D crystals showing a very high thermal conductivity above the limit of bulk graphite [17, 18]. Eventually with the availability of high quality few-layer graphene (FLG), there were experimental studies revealing the gradual change in thermal properties as there is a dimensional crossover from 2D graphene to 3D bulk graphite [19]. These results initiated a vast array of theoretical and experimental research works on study of thermal properties of graphene, graphene nanoribbons (GNR), embedded graphene to name a few and in general aroused huge interest in thermal transport in low dimensional materials. In this work, we provide a review of thermal transport in graphene and in other carbon based materials both bulk and nanostructures. Carbon allotropes exhibit the uniqueness of spanning over a huge range of thermal conductivity values. The bulk of heat in carbon materials is carried by lattice vibrations, i.e. acoustic phonons. While amorphous or disordered carbon has K as low as 0.1 W/mK, single layer graphene and CNT's exhibit values over 3000 W/mK. Diamond and bulk graphite have values of $K \sim$ 2000W/mK at room temperature (RT), the highest amongst bulk materials.

THERMAL CONDUCTIVITY OF BULK GRAPHITE AND DISORDERED CARBONS

In the beginning, we discuss thermal properties of bulk carbon allotropes like diamond and graphite and also thermal transport in disordered carbon materials. These are strikingly different material systems but will help in providing a reference frame to understand thermal properties of graphene and other carbon based nanostructures.

Thermal Properties of Bulk Graphite

In-plane graphite is known to have one of the highest thermal conductivity (~2000 W/mK) amongst carbon based materials [20]. Researchers have investigated c-axis thermal conductivity of graphite using Debye model [21, 22]. These values are smaller than the values extracted for graphite basal planes. The calculations have been done assuming that graphite consists of a few monolayers; defect and boundary scattering have not been taken into considerations. In bulk graphite or highly ordered pyrolytic graphite (HOPG), the crystallites are in perfect alignment with each other and it is almost like a single crystal system. This explains the high in-plane thermal conductivity. However, with the introduction of grain boundaries and improperly oriented grains in polycrystalline graphite, K reduces to 100-300 W/mK at RT due to phonon scattering at these grain boundaries [23]. Acoustic phonons are the primary carrier of heat in these cases. The thermal conductivity values of bulk graphite and diamond exhibit a crystalline characteristic with peaks at 80-100 K and then rolling off with ~1/T dependence. *K* is only limited by Umklapp scattering.

Polycrystalline and Disordered Carbon Materials

Introduction of crystalline grains, point defects and impurities are considered as disorders in a perfect crystalline system and it holds true for carbon based materials as well. These extrinsic factors limit thermal conductivity through changes in phonon scattering, phonon "hopping" mechanisms etc. Amorphous carbon (*a*-C) occupies the lowest thermal conductivity in the carbon-family with 0.01 W/mK at very low temperatures and going up to 1-2 W/mK near RT. In line with other disordered materials, there is a monotonic increase in *K* with T and the main heat conducting mechanism is "hopping" or random walks of localized excitations [24]. Among the disordered carbons, diamond-like-carbon (DLC) is a form of amorphous carbon film with a large percentage of C-C sp^3 bonds and finds applications as protective coatings in micro-electromechanical systems. There is a wide range of DLC materials distinguished in terms of their hydrogen (H) and sp^3 content, degree of disorder and different preparation methods [25]. DLC's which are hydrogen-free and have highest sp^3 content are called tetrahedral amorphous carbon (*ta*-C). Thermal conductivity of these materials essentially depends on the sp^3 content and the structural disorder of the sp^3 phase. Much higher *K*-values ~1.4-3.5 W/mK are shown by *ta*-C samples as compared to DLCH with only ~ 0.7 W/mK [26, 27].

Another interesting set of disordered materials are polycrystalline diamond films which include nanocrystalline diamond (NCD), ultrananocrystalline diamond (UNCD) and microcrystalline diamond (MCD). Thermal conductivity of these films is strongly dependent on the grain size. The average grain size of NCD and UNCD films are respectively 22-26 nm and 2-5 nm. The RT thermal conductivity of UNCD varies in the range 1-10 W/mK. The thermal conductivity of NCD samples with the addition of 25% N_2 is about 9 W/mK while that of the "undoped" films is ~ 16 W/mK. Addition of nitrogen actually creates additional scattering centers for acoustic phonons. In case of MCD, as the name suggests, the grain size is ~ 3-4 μm with K~ 551 W/mK [28, 29]. The main trend observed in these results is the increase of the thermal conductivity with temperature around RT as in fully disordered materials. The specifics of thermal transport in NCD and MCD films can be explained using the phonon-hopping model [30]. These polycrystalline materials are actually partially disordered with general values of *K* lying in between those of bulk crystals and the fully amorphous or disordered materials as shown in Figure 1.

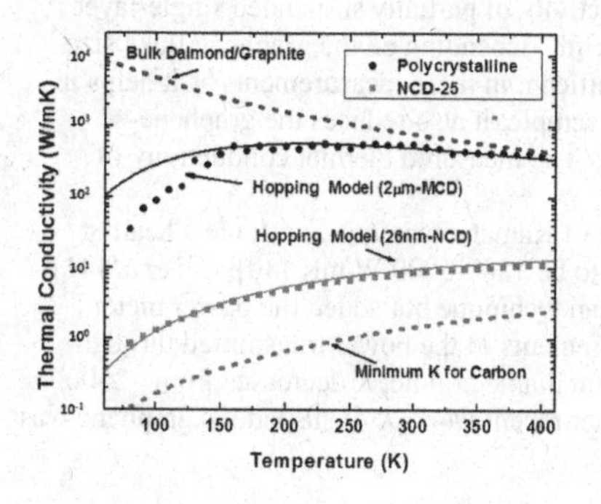

Figure 1: Temperature dependent thermal conductivity values of MCD, NCD, bulk crystals and amorphous carbon materials. The figure is adapted from Refs. [28] and [29].

THERMAL CONDUCTIVITY OF CARBON NANOTUBES

After discovery of carbon nanotubes (CNT), which is a one-dimensional form of carbon [31], there were speculations that this material will have thermal conductivity even higher than crystal graphite [32]. Theoretical studies of thermal conductivity of infinitely long nanotubes have yielded high values comparable to or higher than those of graphite [33]. Single walled carbon nanotubes (SWCNT) grown in crystalline bundles were studied by Hone *et al.* [34]. The tubes were few microns in length with a diameter around 1.4 nm. Room temperature (RT) values of K for single rope of CNTs were found to be 1750-5800 W/mK. Though graphite and CNT, both are made of graphene sheets, yet in the case of highly ordered pyrolytic graphite (HOPG), the *ab*-plane thermal conductivity, dominated by acoustic phonons varies as $T^{2\text{-}3}$ till around 150 K. At higher temperature, the phonon Umklapp scattering process causes the K to decrease with increasing T. The inter-plane vibrations in graphite give rise to extra phonon modes which are absent in single nanotube. The tubular shape of CNT affects the phonon spectrum and scattering times significantly [35]. The thermal conductivity and thermoelectric power of an isolated multi walled CNT (MWCNT) were measured and the value of K was found to be ~ 3000 W/mK around RT [36]. This was subsequently followed by thermal conductance analysis of a suspended metallic SWCNT of 1.7 nm diameter, which gave K values of ~3500 W/mK at RT [37]. A very recent work by Aliev *et al.* [38] shows that the K value for single MWCNT is 600±100 W/mK, bundled MWCNT has 150 W/mK and that of aligned, free standing MWCNT sheet is 50 W/mK. The gradual decrease in K was attributed to the quenching of phonon modes in CNT. A theoretical work by Berber *et al.* [39] used molecular dynamic (MD) simulations to determine thermal conductivity of an isolated nanotube at room temperature and indicates an extremely high value, $K \approx 6600$ W/m K. The same calculations suggested that the thermal conductivity of graphene, a single planar layer of carbon atoms would be even higher.

THERMAL CONDUCTIVITY OF GRAPHENE

Experimental Observations

The first measurements of thermal conductivity of graphene were carried out by Balandin *et al.* [17] and Ghosh *et al.* [18] using the optical Raman spectroscopy as a thermal tool where the local temperature rise due to the laser heating was determined through the independently measured temperature coefficients of the peaks in graphene Raman spectrum [16]. It was found that the near room-temperature (RT) thermal conductivity of partially suspended single-layer graphene (SLG) is in the range K ~ 2000 – 5300 W/mK depending on the graphene flake size.. The suspended portion of the graphene flake is significant in these measurements as it helps in forming an in-plane specific heat wave through the sample; it also reduces the graphene-substrate coupling and substrate-impurity scattering. The measured thermal conductivity of suspended graphene is close to its intrinsic value.

An independent study by using combination of Raman technique and Joule's heating showed the K for CVD grown suspended graphene to be 1500-5000 W/mK [40]. Cai *et al* [41] and Chen *et al.* [42] also used the optothermal Raman technique but added the power meter under the suspended portion of graphene for measurements of the power transmitted through graphene. In case of graphene suspended over 3.8 μm diameter hole, K decreases from ~2500 W/mK at 350 K to ~1400 W/mK at 500 K. In another recent work, K of suspended graphene was found to be 630 W/mK at temperature above RT [43]. For practical applications, it is important to know the thermal conductivity of graphene supported on a substrate. Experimental observations by Seol *et al.* [44] suggest that *K* of single layer graphene exfoliated on a SiO_2 support is ~ 600 W/mK which is still high as compared to metals such as copper. This value is lower than that of freely suspended graphene owing to phonon leakage across graphene-substrate interface and strong interface-scattering of the phonon modes.

Theoretical Studies

These experimental observations motivated theoretical work on the subject. Nika *et al.* [45] performed detailed numerical study of the thermal conductivity of graphene using the phonon dispersion obtained from the valence-force field (VFF) method, and treating the three-phonon Umklapp scattering directly considering all phonon relaxation channels allowed in graphene's 2D Brillouin zone (BZ) [45]. The three- phonon Umklapp processes were treated accounting for all phonon relaxation channels allowed by the momentum and energy conservation laws. The phonon scattering on defects and graphene edges have been also included in the model with mode-dependent Gruneisen parameter, γ taken from ab initio theory. The authors also proposed a simple model for the lattice thermal conductivity of graphene [46] showing that the Umklapp-limited thermal conductivity of graphene increases with the linear dimensions of graphene flakes; averaged values of γ. The results are in agreement with experimental data in [17, 18]. This work showed that the *K* increases with the increase in lateral size or width of the flake; in fact it is a logarithmic divergence and exceeds the *K* for bulk graphite as shown in Figure 2. The figure also shows the experimental set up for the first-ever thermal conductivity measurement.

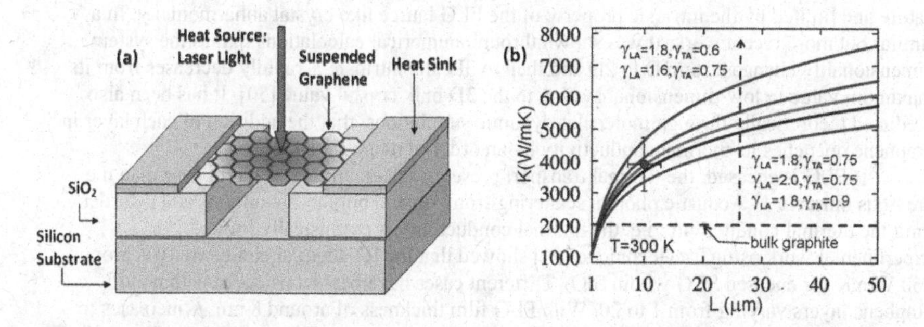

Figure 2: Experimental and theoretical investigation of thermal conduction in SLG. (a) Experimental set up for thermal conductivity measurement of suspended graphene using opto- thermal Raman spectroscopy. (b) Thermal conductivity of graphene as a function of the graphene flake size. It is to be noted that the thermal conductivity of graphene exceeds that of basal planes of graphite when the flake size is larger than few micrometers. The figures are reprinted from Ref. [46] with permission from the American Institute of Physics.

In an effort to understand the difference in thermal conduction in graphene and bulk graphite, we can refer to Klemens' approximation [47, 48]. It explains that, in graphite, thermal transport is 2D only to a frequency of ~ 4 THz and below that it becomes 3D in nature due to cross plane coupling of optical phonons. However, in graphene, the low-energy phonons contribute to thermal conduction all the way to zero frequency and in the integration for Umklapp scattering-limited thermal conductivity, the cut-off frequency is dependent on the phonon mean free path (~775 nm) which is again limited by the graphene flake size. The introduction of grain boundaries, point defects and impurities can provide finite values of K and evade the logarithmic divergence.

THERMAL CONDUCTIVITY OF FEW-LAYER GRAPHENE

The evolution of heat conduction as one goes from 2-D graphene to few-layer graphene (FLG) and, then to 3-D bulk, is of great interest for both fundamental science and practical applications. Thermal conductivity of few-layer graphene (FLG) with varying no. of layers was measured using the Raman spectroscopy technique that has been used for SLG earlier [17, 18]. It was found that the near RT thermal conductivity changes from K~2800 W/mK to 1300 W/mK as the number of atomic plains n increases from n=2 to n=4 [49]. In this case, the changes in K value with n are mostly due to modification of the three-phonon Umklapp scattering and cross-plane phonon coupling of the low-energy phonons. A heat conduction crossover was observed from 2-D graphene to 3-D graphite as seen in Figure 3a. The lower K in n=4 can be explained by stronger boundary scattering effects which result from non uniform thickness of samples The bulk value of K along the basal planes is recovered at $n\approx 8$. As n increases, more phase space becomes available for phonon scattering and the phonon dispersion changes leading to a decrease in K. This thermal conductivity of uncapped suspended FLG is actually intrinsic in nature and limited by the intrinsic property of the FLG lattice like crystal anharmonicity. In a similar but more recent work, it was shown through numerical calculations that as the system dimensionality changes from1D to 2D and then to 3D, the intrinsic K rapidly decreases from its maximum value in low-dimensional system to the 3D bulk crystal value [50]. It has been also predicted theoretically through molecular dynamic simulations, that the addition of each layer in graphene quenches its thermal conductivity by an order of magnitude [36].

If FLG is encased, the thermal transport presents a distinctively different case than the previous intrinsic K. Acoustic phonon scattering from top and bottom boundaries and disorder limit the thermal conductivity; i.e. the thermal conductivity is extrinsically limited. An experimental work using 3-ω technique [51] showed that the RT thermal conductivity is around 160 W/mK for encased SLG within SiO_2. Different cases have been considered with no. of graphene layers varying from 1 to 20. With FLG film thickness of around 8 nm, K increases to ~ 1000 W/mK as shown in Figure 3b. This shows a reverse trend as compared to that of intrinsic K discussed above. In case of encased FLG, there is a strong effect due to disorder penetration through the oxide layers. The K dependence on thickness of FLG is similar to the behavior of encased diamond-like carbon films (DLC) [27]; though the later is an example of amorphous material with disordered sp^2 phase in the interface layers. The average thermal conductivity values of DLC are much smaller than that of FLG but it is interesting to see the similar effects due to boundary and disorder induced phonon scattering.

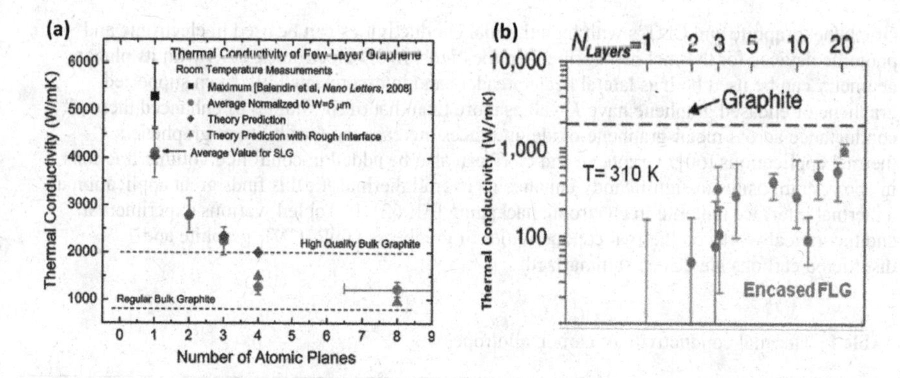

Figure 3: (a) Intrinsic *K* of suspended FLG as a function of number of atomic planes. Figure after Ref [49], reproduced with permission from Nature Publishing Group. (b) Measured *K* of FLG encased in SiO_2 as a function of the thickness of films. Figure adapted from Ref [51] with permission from author and American Chemical Society Publications

THERMAL CONDUCTIVITY OF GRAPHNE NANORIBBONS

With increasing interest in thermal conductivity studies in graphene, there was also a parallel interest in graphene nanoribbons or GNR. These are essentially graphene-strips of width < 20 nm with excellent electrical, mechanical and thermal properties as graphene and potential of applications in electronic industry. Also, as an added advantage they are similar to CNT's yet have a much simpler fabrication process [52]. Lan *et al.* [53] determined the thermal conductivity of graphene nanoribbons combining the tight-binding approach and the phonon non-equilibrium Green's function method. The authors found a thermal conductivity K=3410 W/mK, which is clearly above the bulk graphite limit of 2000 W/mK. A strong edge effect was also revealed by the numerical data. Murali *et al.*, [54] showed that for 20 nm wide graphene nanoribbons the value of K was ~ 1000W/mK. Molecular dynamic (MD) studies [55] gave *K*~2000W/mK for symmetric graphene nanoribbons (GNR). Nanoribbons with zigzag edges had larger *K* than those with armchair edges, and thermal rectification was observed for asymmetric nanoribbons. A strong dependence of thermal conductivity of GNR on shapes and edges was reported by Guo *et al.* [56], similar to the size-dependence reported in [45, 46]. Tensile uniaxial strain was found to distinctly reduce the thermal conductivity of GNR. Equilibrium MD simulations were used to compare thermal conductivity of GNR with smooth and rough edges [57]. A recent work studied the thickness dependent thermal conductivity for few-layer graphene using non-equilibrium MD [58]. Both zigzag and armchair GNR's of width ~1-6 nm have been used for these studies and the *K* values show a dimensional transition from 2D to 3D.

THERMAL MANAGEMENT APPLICATIONS OF CARBON ALLOTROPES

Carbon materials covering a huge range of thermal conductivity values can serve both as thermal insulators, e.g. diamond-like carbon, as well as 'superconductors of heat', such as graphene.

Graphene, graphite and CNT's with high thermal conductivities can be used in electronic and photonic devices for thermal management applications. In 3D electronics, FLG with its planar geometry can be used both as lateral heat spreaders and interconnects [59]. Even supported graphene or encased graphene have *K*-values more than that of Si. Studies of enhanced thermal conductance across metal-graphene-oxide interfaces increase the feasibility of graphene's thermal applications [60]. Graphene and CNT can also be added in controlled volume fractions in epoxy composites to significantly enhance its overall thermal *K*.; this finds great application as a thermal interface material in electronic packaging [61, 62]. In Table I, various experimental and theoretical works on thermal conductivities of graphene, GNR, CNT, graphite and disordered carbons have been summarized.

Table I: Thermal conductivity of carbon allotropes

Sample	K (W/mK)	Method	Comments	Ref
graphene	~ 3080 – 5300	optical	suspended, exfoliated	[17, 18]
FLG	1300-2800	optical	suspended, exfoliated	[49]
graphene	2500	optical	Suspended, CVD grown	[41]
graphene	600	electrical	exfoliated on SiO_2 substrate	[44]
graphene	1500-5000	optical	Suspended, CVD grown	[40]
graphene	2000-5000	Theory VFF	Strong width dependence	[45]
graphene	1-5000	Theory RTA	Strong size dependence	[46]
GNR	3500-6000	MD simulations	100 Å width	[57]
FLG ribbon	1100	electrical	Supported, exfoliated	[54]
MW-CNT	> 3000	electrical	individual	[36]
SW-CNT	~ 3500	electrical	individual	[37]
SW-CNT	1750 – 5800	thermocouples	bundles	[34]
CNT	6600	Theory MD	$K_{CNT} < K_{graphene}$	[39]
graphite	~ 2000	variety	in-plane	[47, 48]
DLCH	~ 0.6 - 0.7	3-omega	H: ~20-35%	[26]
NCD	~ 16	3-omega	grain size: 22 nm	[28]
UNCD	~ 6 - 17	3-omega	grain size: < 26 nm	[29]
ta-C	3.5	3-omega	sp^3: ~ 90%	[26]
ta-C	1.4	3-omega	sp^3 : ~ 60%	[27]

CONCLUSIONS

We have reviewed the results of the experimental and theoretical investigation of heat conduction in graphene, carbon nanotubes, bulk graphite and disordered carbons. The enhanced thermal conductivity of graphene as compared to that of bulk graphite basal planes can be explained by the 2D-nature of thermal transport in graphene over the entire range of phonon frequencies. The superior thermal properties of graphene and other carbon based nanostructures are beneficial for the proposed device applications and thermal management of nanoelectronic chips.

ACKNOWLEDGMENTS

The work in Dr. Balandin's group was supported, in part, by DARPA – SRC through the FCRP Center on Functional Engineered Nano Architectonics (FENA) and Interconnect Focus Center (IFC). The authors are thankful to the current and former members of the Nano-Device Laboratory who contributed to this investigation.

REFERENCES

1. G. E. Moore, *Electronics* **38,** 114 (1965).
2. W. Haensch, E. J. Nowak, R. H. Dennard, P. M. Solomon, A. Bryant, O. H. Dokumaci, A. Kumar, X. Wang, *et al. IBM Journal of Research and Development* **50,** 339 (2006).
3. E. Pop, S. Sinha, and K. E. Goodson, *Proceedings of the IEEE* **94**, 1587 (2006).
4. A. A. Balandin, *IEEE Spectrum*, **29** (2009).
5. D. G. Cahill, W. K. Ford, K. E. Goodson, G. D. Mahan, A. Majumdar, H. J. Maris, R. Merlin, and S. R. Phillpot, *Journal of Applied Physics* **93**, 793 (2003).
6. T, Borca-Tasciuc, D. Achimov, W. L. Liu, G. Chen, H.-W. Ren, C.- H. Lin, and S. S. Pei, *Microscale Thermophys. Eng.* **5**, 225 (2001).
7. A. A. Balandin, and K. L. Wang, *Physical Review B* **58**, 1544 (1998).
8. S. Lepri, R.Livi, and A. Politi, *Physics Reports* **377**, 1(2003).
9. G. Basile, C. Bernardin, and S. Olla, *Physical Review Letters* **96**, 204303 (2006).
10. C. W. Chang, D. Okawa, H. Garcia, A. Majumdar, and A. Zettl, *Physical Review Letters* **101**, 075903 (2008).
11. O. Narayan, and S. Ramaswamy, *Physical Review Letters* **89**, 200601 (2002).
12. K. S. Novoselov, A. K. Geim, S. V. Morozov, D. Jiang, Y. Zhang, S. V. Dubonos, I.V. Grigorieva, and A. A. Firsov, *Science* **306**, 666 (2004).
13. A. K. Geim, and K. S. Novoselov, *Nature Materials* **6**, 183 (2007).
14. K. S. Novoselov, A. K. Geim, S. V. Morozov, D. Jiang, M. I. Katsnelson, I. V. Grigorieva, S.V. Dubonos, and A. A. Firsov, *Nature* **438**, 197 (2005).
15. A. C. Ferrari, J. C. Meyer, V. Scardaci, C. Casiraghi, M. Lazzeri, F. Mauri, S. Piscanec, D. Jiang, K. S. Novoselov, S. Roth, and A. K. Geim, *Physical Review Letters* **97,** 187401 (2006).
16. I. Calizo, W. Bao, F. Miao, C. N. Lau, and A. A. Balandin, *Applied Physics Letters* **91**, 201904 (2007).

17. A. A. Balandin, S. Ghosh, W. Bao, I. Calizo, D. Teweldebrhan, F. Miao, and C. N. Lau, *Nano Letters* **8**, 902 (2008).
18. S. Ghosh, I. Calizo, D. Teweldebrhan, E. P. Pokatilov, D. L. Nika, A. A. Balandin, W. Bao, F. Miao, and C. N. Lau, *Applied Physics Letters* **92**, 151911 (2008).
19. S. Ghosh, W. Bao, D. L. Nika, S. Subrina, E. P. Pokatilov, C. N. Lau, and A. A. Balandin, *Nature Materials* **9**, 555 (2010).
20. B. T. Kelly, *Physics of Graphite*, Applied Science Publishers, London (1986).
21. K. Sun, M. A. Stroscio, and M. Dutta, *Superlattices and Microstructures* **45**, 60 (2009)
22. P. G. Klemens, "Unusually high thermal conductivity in carbon nanotubes," *Proceedings of the Twenty-Sixth International Thermal Conductivity Conference*, ed. R. B. Dinwiddie, (Destech Publications, Lancaster, Pennsylvania, 2004) **26**, pp. 48-57.
23. A. L. Woodcraft, M. Barucci, P. R. Hastings, L. Lolli, V. Martelli, L. Risegari, and G. Ventura, *Cryogenics* **49**, 159 (2009).
24. D. G, Cahill, and R. O. Pohl, *Solid State Communications*, **70**, 927 (1989).
25. C. Casiraghi, A. C. Ferrari, and J. Robertson, *Physical Review B* **72**, 085401 (2005).
26. M. Shamsa, W. L. Liu, A. A. Balandin, C. Casiraghi, W. I. Milne, and A. C. Ferrari, *Applied Physics Letters* **89**, 161921 (2006).
27. A. A. Balandin, M. Shamsa, W. L. Liu, C. Casiraghi, and A. C. Ferrari, *Applied Physics Letters* **93**, 043115 (2008).
28. W. L. Liu, M. Shamsa, I. Calizo, A. A. Balandin, V. Ralchenko, A. Popovich, and A. Saveliev, *Applied Physics Letters* **89**, 171915 (2006).
29. M. Shamsa, S. Ghosh, I. Calizo, V. Ralchenko, A. Popovich, and A. A. Balandin, *Journal of Applied Physics* **103**, 083538 (2008).
30. L. Braginsky, V. Shklover, H. Hofmann, and P. Bowen, *Physical Review B* **70**, 134201 (2004).
31. S. Ijima, *Nature* **354**, 56 (1991).
32. R. S. Ruoff, and D. C. Lorents, *Carbon* **33**, 925 (1995).
33. M. A. Osman, and D. Srivastava, *Nanotechnolgy* **12**, 21 (2001).
34. J. Hone, M.Whitney, C. Piskoti, and A. Zettl, *Physical Review B* **59**, R2514 (1999).
35. L. X. Benedict, S. G. Louie, and M. L. Cohen, *Solid State Communications* **100**, 177 (1996).
36. P. Kim, L. Shi, A. Majumder, P. L. McEuen, *Physical Review Letters* **87**, 215502 (2001).
37. E. Pop, D. Mann, Q. Wang, K. Goodson, and H. Dai, *Nano Letters* **6**, 96 (2006).
38. A. E. Aliev, M. H. Lima, E. M. Silverman, and R. H. Baughman, *Nanotechnology* **21**, 035709 (2010).
39. S. Berber, Y-K. Kwon, and D. Tomanek. *Physical Review Letters* **84**, 4613 (2000).
40. L. A. Jauregui, Y. Yue, A. N. Sidorov, J. Hu, Q. Yu, G. Lopez, R. Jalilian, D. K. Benjamin, et al. , *Electrochemical Society Transactions* **28**, 73 (2010).
41. W. Cai, A. L. Moore, Y. Zhu, X. Li, S. Chen, L. Shi, and R. S. Ruoff, *Nano Letters* **10**, 1645 (2010).
42. S. Chen, A. L. Moore, W. Cai, J. W. Suk, J. An, C. Mishra, C. Amos, C. W. Magnuson, J. Kang, L. Shi, and R. S. Ruoff, *ACS Nano* **5**, 321 (2011).
43. C. Faugeras, B. Faugeras, M. Orlita, M. Potemski, R. R. Nair, and A. K. Geim, *ACS Nano* **4**, 1889 (2010).
44. J. H. Seol, I. Jo, A. R, Moore, L. Lindsay, Z. H. Aitken, M. T. Pettes, X. Li, Z. Yao, R. Huang, D. Broido, N. Mingo, and R. S. Ruoff, *Science* **328**, 213 (2010).

45. D. L. Nika, E. P. Pokatilov, A. S. Askerov, and A. A. Balandin, *Physical Review B* **79**, 155413 (2009).
46. D. L. Nika, S. Ghosh, E. P. Pokatilov, and A. A. Balandin, *Applied Physics Letters* **94**, 203103 (2009).
47. P. G. Klemens, *J. Wide Bandgap Materials* **7**, 332 (2000).
48. P. G. Klemens, *Int. J. Thermophysics* **22**, 265 (2001).
49. S. Ghosh, W. Bao, D. L. Nika, S. Subrina, E. P. Pokatilov, C. N. Lau, and A. A. Balandin, *Nature Materials* **9**, 555 (2010).
50. K. Saito, and A. Dhar, *Physical Review Letters* **104**, 040601 (2010).
51. W. Jang, Z. Chen, W. Bao, C. N. Lau, and C. Dames, *Nano Letters* **10**, 3909 (2010).
52. A. Naeemi, and J. D. Meindl, *IEEE Electron Device Letters* **28**, 428 (2007).
53. J. Lan, J. S. Wang, C. K. Gan, and S. K. Chin, *Physical Review B* **79**, 115401 (2009).
54. R. Murali, Y. Yang, K. Brenner, T. Beck, and J. D. Meindl, *Applied Physics Letters* **94**, 243114 (2009).
55. J. Hu, X. Ruan, and Y. P. Chen, *Nano Letters* **9**, 2730 (2009).
56. Z. Guo, D. Zhang, and X-G. Gong, *Applied Physics Letters* **95**, 163103 (2009).
57. W. J. Evans, L. Hu, and P. Keblinski, *Applied Physics Letters* **96**, 203103 (2010).
58. W.-R. Zhong, M. P. Zhang, B. Q. Ai, and D. Q. Zheng, *Applied Physics Letters* **98**, 113107 (2011).
59. S. Subrina, D. Kotchekov, and A. A. Balandin, *IEEE Electron Device Letters* **30**, 1281 (2009).
60. Y. K. Koh, M. H. Bae, D. G. Cahill, and E. Pop, *Nano Letters* **10**, 4363 (2010).
61. K. M. F. Sahil, V. Goyal, and A. A. Balandin, *ECS Proceeds*, (2011).
62. A. Yu, M. E. Itkis, E. Bekyarova, and R. C. Haddon, *Applied Physics Letters* **89**, 133102 (2006).

Mater. Res. Soc. Symp. Proc. Vol. 1344 © 2011 Materials Research Society
DOI: 10.1557/opl.2011.1348

Experimental Demonstration of Thermal Management of High-Power GaN Transistors with Graphene Lateral Heat Spreaders

Zhong Yan, Guanxiong Liu, Javed Khan, Jie Yu, Samia Subrina and Alexander Balandin

Nano-Device Laboratory, Department of Electrical Engineering and Materials Science and Engineering Program, University of California - Riverside, Riverside, California 92521 USA

ABSTRACT

Graphene is a promising candidate material for thermal management of high-power electronics owing to its high intrinsic thermal conductivity. Here we report preliminary results of the proof-of-concept demonstration of graphene lateral heat spreaders. Graphene flakes were transferred on top of GaN devices through the mechanical exfoliation method. The temperature rise in the GaN device channels was monitored *in-situ* using micro-Raman spectroscopy. The local temperature was measured from the shift in the Raman peak positions. By comparing Raman spectra of GaN devices with and without graphene heat spreader, we demonstrated that graphene lateral heat spreaders effectively reduced the local temperature by ~ 20°C for a given dissipated power density. Numerical simulation of heat dissipation in the considered device structures gave results consistent with the experimental data.

INTRODUCTION

Self-heating is a severe problem for high-power GaN electronics. A possible method for improving heat removal from GaN/AlGaN heterostructure field-effect transistors (HFETs) is introduction of an additional heat escape channel using materials with high thermal conductivity. Graphene is a promising candidate material for heat removal and thermal management applications. In 2008, our group discovered that the intrinsic thermal conductivity of single-layer graphene is extremely high and can exceed that of bulk graphite [1]. The defects and coupling to the substrate reduce thermal conductivity of graphene. However, the bulk graphite limit, K ~2000 W/mK, achieved in few-layer graphene (FLG) of certain thickness [2] is still much higher than thermal conductivity of conventional semiconductors or metals (e.g. K~145 W/mK for bulk crystalline Si or K~400 W/mK for bulk copper).

In this study, we report preliminary results, which suggest a possible use of graphene flakes for heat removal. For the proof-of-concept demonstration, mechanical exfoliated graphene flakes were transferred on top of GaN/AlGaN HFETs. We measured the temperature rise in GaN HFETs using a micro-Raman spectrometer. The Raman peak position in the spectra of GaN depends on temperature [3]. Knowing the temperature coefficients of the Raman peaks we could monitor the temperature rise from the shifts in the position of Raman peaks. By comparing the Raman spectra for the GaN HFETs with and without graphene heat spreaders on top, it was demonstrated that graphene layers reduced the local temperature by ~20 °C for the power densities involved. The numerical simulations of heat dissipation in the considered device structures were carried out with the help of COMSOL software and the results were consistent with the experimental data. The computer modeling approach was similar to the one reported earlier for the graphene-SOI structures [4].

EXPERIMENTAL APPROACH

The devices used for this study - AlGaN/GaN HFETs – had the top gate metal contact isolated from AlGaN barrier layer by a thin SiO_2 film. The device consisted of 30 nm AlGaN (~20% Al) on 0.5 μm GaN layer deposited on an insulating 4H-SiC substrate. The Ohmic contacts for the source and drain were fabricated by depositing Ti/Al/Ti/Au [5-6]. The graphene flakes were produced by a standard mechanical exfoliation method from the highly oriented pyrolytic graphite (HOPG). In order to transfer graphene flakes on top of GaN HFETs, we first tried to exfoliate graphene flaks directly on a diced chip with GaN devices on top. After repeating mechanical exfoliation process on a same chip for several times, a number of graphene flakes with different size and thickness were transferred to that chip. The main challenge of this method was the random location of graphene flakes. As an alternative approach we used the PMMA transfer method, which enables the precise control of the location of graphene flakes.

The *in-situ* monitoring of the temperature rise at the hot spots of GaN HFETs was carried out using micro-Raman spectroscopy. Micro-Raman spectroscopy has been widely used in the investigation of device temperature in AlGaN/GaN HFETs [7-8]. It provides a nondestructive, high-throughput and high-spatial-resolution method to study local temperature rises in electronic devices. The measurements were performed on GaN HFETs in $z(x,x)\bar{z}$ backscattering geometry using a Renishaw Raman system with the 488 nm laser as an excitation source. The E_2 phonon mode of GaN was used to extract the temperature rise. The laser excitation power was substantially lower than the electrical power applied to GaN HFETs during the Raman measurements. The Raman peak position was identified by fitting the experimental data with the Lorentzian function to increase the spectrum resolution to about 0.1 cm^{-1}.

RESULTS AND DISCUSSION

Figure 1 shows an optical image (a) and DC-IV characteristics (b) of a typical tested GaN HFET. In this double gate-finger HFET device the gate length and width are 3.5 μm and 90 μm, respectively. The device was completely pinched off at negative gate bias of -4 V. The negative slope regions in the I-V curves indicate a degradation of the carrier mobility and increase of the channel resistance due to Joule heating when dissipated power increases. We also measured the DC IV characteristics of this GaN HFET at increased ambient temperature by external heating through a hot disk to verify that the source-drain current density was very sensitive to temperature. The current density decreases approximately linear with increasing temperature in the tested temperature range.

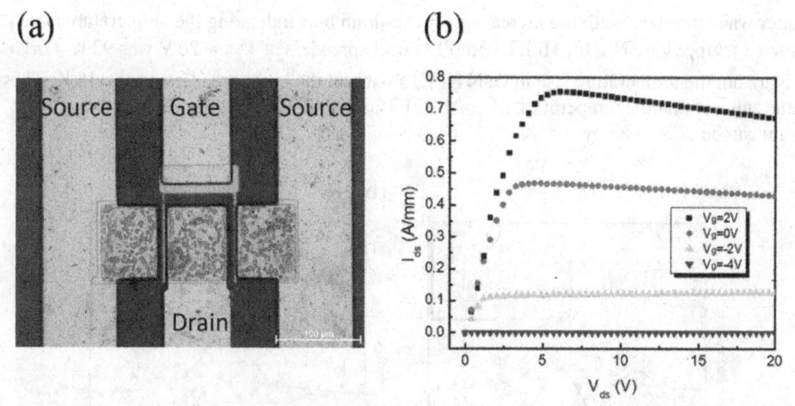

Figure 1: (a) Optical image of the tested two-finger GaN HFET device. The gate width is 90 μm. The scale bar is 100 μm. (b) DC I-V characteristics of the tested GaN HFET device. The gate bias varied from 2 V to -4 V at -2 V intervals. The drain bias was scanned from 0 V to 20 V.

Figure 2 shows the schematics of the few-layer graphene (FLG) flakes transferred on top of GaN HFET device. Special care was taken to provide a good thermal contact while avoiding shortening the source and gate.

Figure 2: Schematic of graphene lateral heat spreader transferred on top of a high-power GaN HFET device.

The results of Raman measurements of temperature rise in GaN HFETs with and without graphene lateral heat spreaders are shown in Figure 3. The Raman peak position shifts to the smaller wave numbers with the increasing source-drain bias indicating the temperature rise. The extracted temperature rise for HFET with FLG heat spreaders at $V_{DS} \approx 20$ V was 92 K. For comparison, the temperature rise in GaN HFETs without the heat spreaders was ~118 K. These results suggest that the temperature of GaN HFET at the dissipated power density of ~12.8 W/mm can be decreased by ~20 K.

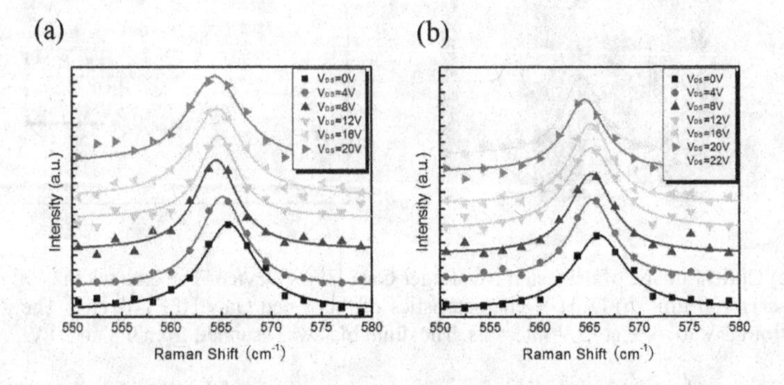

Figure 3: Micro-Raman *in situ* monitoring of temperature rise in GaN HFETs with (a) and without (b) graphene lateral heat spreader. The solid lines are the Lorentzian function fitting curves to the experimental data. The temperature is determined from the spectral position of the E_2 phonon mode of GaN.

Three-dimensional numerical simulations of heat dissipation in the considered GaN HFET were carried out with finite-element method using COMSOL software. The modeling results are consistent with the experimental data in the estimated error range.

CONCLUSIONS

We have experimentally demonstrated effective thermal management of high-power GaN HFET with the graphene lateral heat spreaders. Temperature rise in the hot spots of the device active channels was monitored with the micro-Raman spectroscopy. The experimental results show that the graphene heat spreaders reduced the device temperature by ~20 K for a given dissipated power density.

ACKNOWLEDGMENTS

This work was supported by ONR through award N00014-10-1-0224 on graphene heat spreaders. The authors thank Dr. S. Rumyantsev and Prof. M. Shur for GaN devices provided for this study.

REFERENCES

1. A. Balandin, S. Goash, W. Z. Bao, I. Calizo, D. Teweldebrhan. F. Miao and C. N. Lau, *Nano Lett.* **8,** 902 (2008)
2. S. Ghosh, W. Z. Bao, D. L. Nika, S. Subrina, E. P. Pokatilov, C. N. Lau and A. A. Balandin, *Nature Materials* **9,** 555 (2010)
3. D. J. Chen, B. Shen, X. L. Wu, J. C. Shen, F. J. Xu, K. X. Zhang, R. Zhang, R. L. Jiang, Y. Shi and Y. D .Zheng, *Appl. Phys. A* **80,** 1729 (2005)
4. S. Subrina, D. Kotchetkov and A. A. Balandin, *Electron Dev. Lett.*, **30**, 1281(2009)
5. M. A. Khan, X. Hu, A. Tarakji, G. Simin, J. Yang, R. Gaska and M. S. Shur, *Appl. Phy. Lett.* **77,** 1339 (2000)
6. G. Simin, X. Hu, N. Ilinskaya, J. Zhang, A. Tarakji, A. Kumar, J. Yang, M. Asif Khan, R. Gaska and M. S. Shur, *IEEE Electron. Dev. Lett.* **22,** 53 (2001)
7. M. Kuball, J. M. Hayes, M, J. Uren, T. Martin, J. C. H. Birbeck, R. S. Balmer and B. T. Hughes, *IEEE Electron. Dev. Lett.* **23,** 7 (2002)
8. A. Sarua, H. Ji, M. Kuball, M. J. Uren, T. Martin, K. P. Hilton and R. S. Balmer, *IEEE Trans. Electron Dev.* **53,** 2438 (2006)

Mater. Res. Soc. Symp. Proc. Vol. 1344 © 2011 Materials Research Society
DOI: 10.1557/opl.2011.1350

Top-Gate Graphene-on-UNCD Transistors with Enhanced Performance

Jie Yu[1], Guanxiong Liu[1], Anirudha V. Sumant[2] and Alexander A. Balandin[1]

[1]Nano-Device Laboratory, Department of Electrical Engineering and Materials Science and Engineering Program, University of California, Riverside, California 92521 USA

[2]Center for Nanoscale Materials, Argonne National Laboratory, IL, 60439 USA

ABSTRACT

We fabricated a number of top-gate graphene field-effect transistors on the ultrananocrystalline diamond (UNCD) – Si composite substrates. Raman spectroscopy, scanning electron microscopy and atomic force microscopy were used to verify the quality of UNCD and graphene device channels. The thermal measurements were carried out with the "hot disk" and "laser flash" methods. It was found that graphene on UNCD devices have increased breakdown current density by ~50% compared to the reference devices fabricated on Si/SiO_2. The relatively smooth surface of UNCD, as compared to other synthetic diamond films, allowed us to fabricate top gate graphene devices with the drift mobility of up to ~ 2587 $cm^2V^{-1}s^{-1}$.

INTRODUCTION

In order to improve the electrical properties and thermal management of graphene devices we examined a possibility of graphene device fabrication on synthetic diamond. It is expected that replacement of silicon oxide with synthetic diamond can improve heat conduction and reduce leakage current in the substrate. We studied graphene on several composite substrates including ultrananocrystalline (UNCD), microcrystalline diamond (MCD) thin films on silicon. The diamond thin films were grown by the microwave plasma chemical vapor deposition (MPCVD) at Argon National Laboratory and characterized using optical microscopy, Raman spectroscopy, scanning electron microscopy (SEM) and atomic force microscopy (AFM). The SEM and TEM inspection revealed that the average grain size of UNCD films is in the range 3 - 5 nm. The UNCD films were polished using CMP process and the average surface roughness for UNCD is on the order of 1 nm measured by AFM. The graphene layers were prepared by the mechanical exfoliation from the bulk highly oriented pyrolytic graphite. The quality of graphene was verified using the micro-Raman spectroscopy by analyzing the G peak and 2D band. In this paper, we discuss the issues of resolving graphene peaks on the Raman spectrum background from UNCD. After the number of layers and quality of graphene-on-diamond were confirmed, the fabrication steps were implemented [1].

The current-voltage (I-V) characteristics of the resulting graphene-on-diamond devices were measured using the semiconductor parameter analyzer. The obtained devices had the carrier mobility ~ 2354 $cm^2V^{-1}S^{-1}$ for holes and ~1293 $cm^2V^{-1}S^{-1}$ for electrons. The breakdown current density of the graphene-on-diamond devices was significantly improved as compared to graphene-on-oxide. It is expected that further improvements in synthetic diamond growth technology will lead to higher mobility and breakdown current density in graphene-on-diamond FETs.

EXPERIMENTAL DETAILS

After graphene is exfoliated the cross-shaped metal landmarks were used for the location of graphene flakes. They were formed using lithography followed by metal deposition by the electron beam evaporator. The 10-nm Ti and 100-nm Au were deposited for the contacts. The electron beam lithography was used to define the regions for the top gate oxide. The electron beam lithography was used to define the source, drain, and the top gate, the gate dielectric was deposited by the low temperature atomic layer deposition. The thickness of HfO_2 deposited directly in the region was 20 nm. Because of the minor conductivity of UNCD substrate, we use the whole HfO_2 under gate metal to separate the gate, source and drain pad for reducing the device leakage, shows a schematic of the top gate graphene transistor structure. We use EBL defined the source, drain, and the top gate, and followed by the electron beam evaporator to make Ti/Au (10/60nm) electrodes and pads.

Figure 1 shows an SEM image of typical device. After fabricated the top gated device, we duplicated the same device structure for each device without oxide and graphene on the same substrate following the same fabrication procedure and parameters to study the effect of the diamond substrate.

Figure 1: Schematic of the graphene-on-diamond top-gate FET (upper panel). The HfO_2 dielectric is indicated by the green color, metal gate and contacts are shown with the yellow color. SEM image of the actual fabricated graphene-on-diamond device is shown in the lower panel. One can see the graphene channel under the HfO_2 gate dielectric. The width of the dielectric layer is 3 μm.

RESULTS AND DISCUSSION

Graphene is sp^2 bonded and it is known that UNCD films have small fraction of sp^2 bonds [2], at the grain boundaries, which complicates Raman inspection of graphene on UNCD. We were still able to identify graphene and few-layer graphene (FLG) via their Raman signatures (Figure 2). After confirming the quality and the number of atomic planes in FLG we proceeded with device fabrication. The action of the top gate of a typical graphene-on-UNCD field-effect transistor is shown in Figure 3.

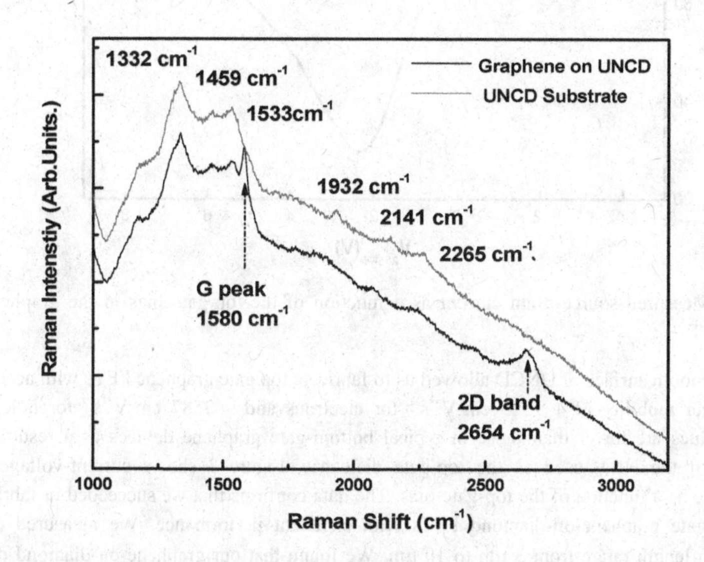

Figure 2: Raman spectrum of FLG on UNCD under 488 nm excitation. For comparison, the spectrum of the UNCD substrate is also shown.

Table I: Raman peak assignment of graphene on UNCD

Measured (cm^{-1})	Explanation		
1140-70	Trans-polyacetylene at the grain boundary [4]		
1332	diamond	1332±0.5	2452±72
1459	1467, a weak feature observed in diamonds [5]		
1533	Amorphous carbon [6]		
2265	$L(X^{(1)})+TO(X^{(4)})$ [7]		

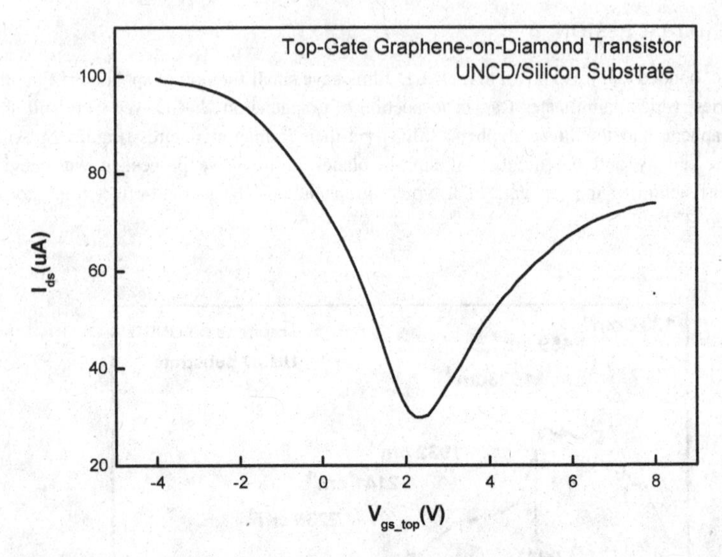

Figure 3: Measured source-drain current as a function of the top-gate bias in the graphene-on-diamond FET.

The smooth surface of UNCD allowed us to fabricate top gate graphene FETs with acceptable charge carrier mobility of ~ 1515 $cm^2V^{-1}s^{-1}$ for electrons and ~ 2587 $cm^2V^{-1}s^{-1}$ for holes. The mobility values are lower than those in typical bottom-gate graphene devices as a result of the deposition of the HfO_2 used as the top-gate dielectric. Figure 3 shows current-voltage (I-V) characteristic as a function of the top-gate bias. The data confirms that we succeeded in fabrication of the top-gate graphene-on-diamond FETs with excellent performance. We measured our 30 devices with length range from 3 μm to 10 μm. We found that our graphene-on-diamond devices have increased breakdown current density by as much as ~50% compared to the reference devices fabricated on Si/SiO_2 and to the typical values reported in literature. Our modelling results obtained with the help of COMSOL simulation tool gave consistent values. In graphene, the failure likely occurs at a certain threshold power through the rapid oxidation in air similarly to that in carbon nanotubes [3]. It was concluded that the primary factor in the breakdown initiation was current-induced defect formation because the oxidation of defect-free graphite can proceed only at extremely high temperatures.

CONCLUSIONS

We have shown that graphene-on-diamond devices with better heat dissipation can withstand remarkable current densities, exceeding 10^9 A/cm^2. Although some technological challenges with synthetic diamond sill remain, our results suggest that graphene-on-diamond can become a promising alternative to graphene-on-silicon-oxide for device applications.

ACKNOWLEDGEMENTS

The work at UCR was supported, in part, by DARPA – SRC Center on Functional Engineered Nano Architectonics (FENA) and DARPA Defense Microelectronics Activity (DMEA) under agreement number H94003-10-2-1003. Use of the Center for Nanoscale Materials was supported by the U. S. Department of Energy, Office of Science, Office of Basic Energy Sciences, under Contract No. DE-AC02-06CH11357.

REFERENCES

[1] G. Liu, W. Stillman, S. Rumyantsev, Q. Shao, M. Shur and A.A. Balandin, "Low-frequency electronic noise in the double-gate single-layer graphene transistors," *App. Phy. Lett.,* **95**, 033103 (2009).

[2] J. E. Butler, A. V. Sumant, "The CVD of nanodiamond materials," *Chem. Vap. Deposition.*, **14**, 145 (2008).

[3] P.G. Collins, M. Arnold, M. Hersam, R. Martel, Ph. Avouris, "Current Saturation and Electrical Breakdown in Multiwalled carbon nanotubes," *Phys. Rev. Lett.,* **86**, 3128 (2001).

[4] A. C. Ferraria, J. Robertson "Origin of the 1150-cm^{-1} Raman mode in nanocrystalline diamond," *Phys. Rev. B,* **63**, 121405R (2001).

[5] A. M. Zaitsev, *Optical Properties of Diamond: Data Handbook (*Springer, 2001).

[6] V.Y. Osipov, A.V. Baranovb, V.A. Ermakovb, T.L. Makarovaa, L.F. Chungongc, A.I. Shamesd, K. Takaie, T. Enokie, Y. Kaburagif, M. Endog and A.Ya. Vul, " Raman characterization and UV optical absorption studies of surface plasmon resonance in multishell nanographite," *Diamond & Related Materials.,* **20**, 205 (2011).

[7] S. A. Solin, A. K. Ramdas, "Raman Spectrum of Diamond," *Phys. Rev. Lett.,* **1**, 1687 (1970).

Mater. Res. Soc. Symp. Proc. Vol. 1344 © 2011 Materials Research Society
DOI: 10.1557/opl.2011.1368

All-Carbon Composite for Photovoltaics

Alvin T.L. Tan, Vincent C. Tung, Jaemyung Kim, Jen-Hsien Huang, Ian Tevis, Chih-Wei Chu, Samuel I. Stupp and Jiaxing Huang

Department of Materials Science and Engineering, Northwestern University, Evanston, USA

ABSTRACT

Graphitic nanomaterials such as graphene, carbon nanotubes (CNT), and C_{60} fullerenes are promising materials for energy applications because of their extraordinary electrical and optical properties. However, graphitic materials are not readily dispersible in water. Strategies to fabricate all-carbon nanocomposites typically involve covalent linking or surface functionalization, which breaks the conjugated electronic networks or contaminates functional carbon surfaces. Here, we demonstrate a facile surfactant-free strategy to create such all-carbon composites. Fullerenes, unfunctionalized single walled carbon nanotubes, and graphene oxide sheets can be conveniently co-assembled in water, resulting in a stable colloidal dispersion amenable to thin film processing. The thin film composite can be made conductive by mild thermal heating. Photovoltaic devices fabricated using the all-carbon composite as the active layer demonstrated an on-off ratio of nearly 10^6, an open circuit voltage of 0.59V, and a power conversion efficiency of 0.21%. This photoconductive and photovoltaic response is unprecedented among all-carbon based materials. Therefore, this surfactant-free, aqueous based approach to making all-carbon composites is promising for applications in optoelectronic devices.

INTRODUCTION

Graphitic nanostructures such as carbon nanotubes and fullerenes (C_{60}) can be used to facilitate separation, transport and collection of charge carriers in organic photovoltaic devices.[1-8] Fullerenes are highly electron accepting and behave as n-type semiconductors.[9] On the other hand, semiconducting carbon nanotubes[10,11] and graphene-based sheets[12,13] exhibit p-type-like behavior in ambient conditions. Therefore, nanoscopic heterojunctions can be generated if these carbon nanostructures are brought together.[1-7,14] Such an approach employing all carbon nanostructures also combines the unique properties of the constituents. Fullerenes are known to exhibit ultra-fast electron transfer[9], while semiconducting carbon nanotubes have high absorption in the near IR range and exhibit extraordinary charge carrier mobility.[10,11] Therefore, a hybrid material of C_{60} and carbon nanotubes can be used as a photovoltaic material with good absorption, good charge separation, and efficient carrier transport.

However, CNT and C_{60} nanomaterials tend to aggregate in aqueous solution. To overcome their poor solubility, fabrication of all-carbon nanocomposites typically involves covalently linking of the constituents, extensive surface functionalization, or the use of non-conducting surfactants.[1-7] These strategies either break the conjugated carbon networks or contaminate the functional heterojunction interfaces. Recently, we showed that graphene oxide (GO), the chemical exfoliation product of graphite, can be used as a surfactant sheet to disperse graphitic materials in water.[13-17] Since GO can be cleanly converted to conductive reduced GO (r-GO),[18-20] electrically addressable carbon-carbon interfaces can be generated. Here, we demonstrate the fabrication of all-carbon composites made of C_{60}, single-walled carbon nanotubes (SWCNT), and graphene. Direct mixing of GO with C_{60} and SWCNT yields a stable colloidal dispersion which is readily solution processable. The composite was incorporated as the active layer of a photovoltaic device, which exhibited an unprecedented photoconductive and photovoltaic response compared to all-carbon composites prepared by other methods.

EXPERIMENTAL SECTION

Graphite powder (SP-1 grade) was purchased from Bay Carbon Inc. Single wall carbon nanotubes (P2-SWCNTs) were purchased from Carbon Solutions, Inc. GO was synthesized by a modified Hummers' method[21] as reported elsewhere.[22] The dispersion was extensively washed and filtered to remove salt by-products and excess acid.[23] The dry GO filter cakes were re-dispersed in water to create a stock solution of 1 mg/ml, which may be further diluted to various concentrations. To create the dispersion of the binary mixtures of C_{60} (Nano C), unfunctionalized SWCNT powders or GO, the individual components are blended in a mixture of water and methanol (10 to 1, v/v) and sonicated using a tip-sonicator (Misonix Ultrasonic Cell Disruptors) for 2 hrs. Typically the concentrations of C_{60}, SWCNTs and GO are 0.5 mg/ml, 1 mg/ml and 1 mg/ml, respectively. To create the ternary composite, the 3 components were directly mixed and sonicated, or alternatively, C_{60} and SWCNT were first sonicated together to form a complex and then added into GO dispersion. Black colloidal dispersions with no visible precipitation were formed by either method. However, the dispersions made by the two-step mixing were found to produce more uniform thin films, and therefore was used for preparing thin films for device studies. Typically to prepare the C_{60}/SWCNTs/GO hybrid, 100 µl of C_{60}/SWCNTs stock solution was carefully added into the GO solution (1 mg/ml) and was sonicated for two hour to ensure the stable colloidal dispersion. The resulting dispersion is stable for months.

Microscopic observations were made with SEM (Hitachi S-4800-II FEG), TEM (JEOL JEM-2100F) and AFM (Multimode, DI). Confocal Raman spectroscopy studies were done at a fixed excitation wavelength of 514.5 nm with thin films spin-casted on a silicon substrate. UV-Vis spectra were acquired with an Agilent 8653 spectrometer. The spectra of SWCNTs and C_{60} were measured on the supernatant right after sonication. As for preparation of all-carbon composite devices, 5-8 nm thick active layer material of C_{60}/SWCNTs/GO was spin coated onto a pre-patterned ITO anode (20 Ω/sq) with a thin buffer layer of PEDOT: PSS (20nm). The layer was annealed in ambient conditions at 150 °C for 30 min to reduce GO. A 15 nm thick corresponding fullerene layer was thermally evaporated followed by a 100 nm thick Al as the top electrode through a shadow mask. Photovoltaic measurements were done under AM1.5 80 $mWcm^{-2}$ illumination from a Thermal Oriel Xe solar simulator equipped with an Oriel 130 monochromator. Filters were used to cut off grating overtones. A calibrated silicon reference solar cell with a KG5 filter certified by the National Renewable Energy Laboratory (NREL) was used to confirm the measurement conditions. Device configurations for external quantum efficiency (EQE) measurement were similar to those used in photovoltaic measurement. The work function measurements were carried out by ultraviolet photoelectron spectroscopy (UPS) (ULVAC-PHI, Chigasaki) under ultrahigh vacuum condition (10^{-8} Pa). UPS spectra were collected on r-GO and SWCNTs thin films spin coated on the silver coated silicon substrates under positive bias. Films were annealed at 150 °C in ambient conditions prior to characterization. Helium lamp emitting at 21.2 eV was used as the light source to excite the valence electrons.

RESULTS AND DISCUSSION

Synergistic Assembly and Characterization of All-Carbon Composites

To create a binary dispersion, unfunctionalized SWCNT and C_{60} were mixed in a water/methanol solution and sonicated. A temporary dispersion of SWCNT and C_{60} was created, as evident by the reduced amount of powdery C_{60} precipitates and darkened supernatant. However, the C_{60}/SWCNT complex was not stable and aggregated rapidly (Figure 1a). The C_{60}/SWCNT complex was not solution processable. Spin-coated films of C_{60}/SWCNT were very uneven (Figure 1b) and profilometer measurements showed that the roughness was on the micrometer scale (Figure 1c). To create a C_{60}/SWCNT/GO ternary dispersion, GO was added to the C_{60}/SWCNT complex and the mixture was sonicated, resulting in a stable colloidal dispersion (Figure 1d). Since GO has unoxidized graphitic domains,[24] the C_{60}/SWCNT complex adheres to the graphitic domains of GO. Meanwhile, the ionized edges of the GO sheet allow the C_{60}/SWCNT/GO complex to stay dispersed in water. The dispersion was processable and can be readily spin coated to constitute uniform thin films (Figure 1e). Surface profile measurements established by AFM showed that roughness was on the nanometer scale (Figure 1f).

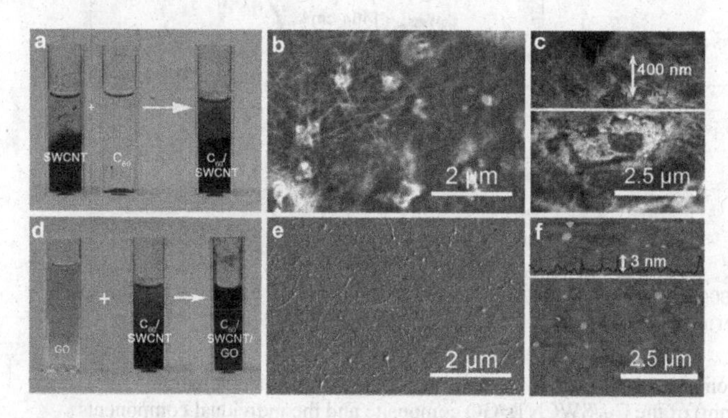

Figure 1. Assembly of all-carbon composites. (a) Mixing SWCNT and C_{60} results in an unstable C_{60}/SWCNT dispersion. (b) SEM of a C_{60}/SWCNT film formed by spin-coating. (c) Optical image and height profile obtained using a profilometer. (d) Addition of GO to the C_{60}/SWCNT mixture results in a stable C_{60}/SWCNT/GO dispersion. (e) SEM of a C_{60}/SWCNT/GO film formed by spin-coating. (f) AFM image and height profile of theC_{60}/SWCNT/GO film.

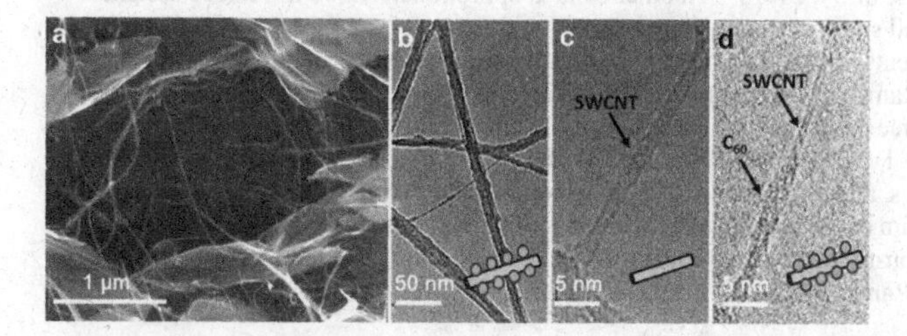

Figure 2. (a) Cross-sectional SEM image of a thick composite film showing SWCNTs bridging exfoliated r-GO sheets. (b) Low resolution TEM image of C_{60}/SWCNT revealing rough coatings on the SWCNT. (c) High resolution TEM image of an uncoated SWCNT. (d) High resolution TEM image of an SWCNT half-coated with clusters of C_{60}.

The GO sheets in the film can be converted to chemically modified graphene via mild thermal treatment, resulting in a C_{60}/SWCNT/r-GO all-carbon composite. Figure 2a shows a cross-sectional SEM of a thick C_{60}/SWCNT/r-GO film. SWCNT are seen bridging the r-GO sheets. Further, TEM imaging shows that the tubes are largely debundled, and high resolution TEM imaging reveals a rough coating on the surface of the SWCNT, which is presumably a layer of C_{60} clusters. Comparing the thickness of uncoated (Figure 2c) and half-coated tubes (Figure 2d), the thickness of the coating was estimated to be 1 to 2 nm.

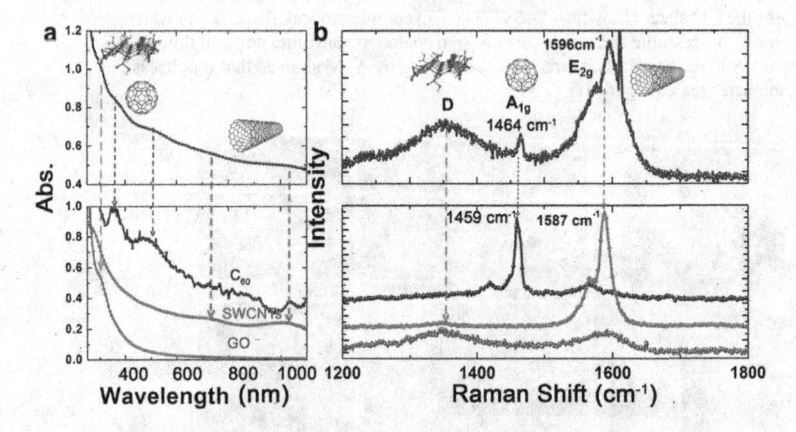

Figure 3. (a) Solution phase UV–Vis and (b) solid state Raman spectra (excitation wavelength of 514 nm) of the C_{60}/SWCNTs/GO composite and the individual components. The red-shift in both spectra indicates strong π-π stacking between the components.

The electronic structures of the constituents in the composites are largely preserved, as is evident by UV-vis spectrum (Figure 3a). C_{60}/SWCNT/GO hybrid dispersion shows all the characteristic absorption bands of the constituents. Two distinct absorption bands at 360 nm and 505 nm can be attributed to C_{60},[25] while two bands at approximately 750 nm and 930 nm are characteristic of SWCNTs.[26] Another band at approximately 300 nm can be assigned to GO.[27] The overall spectrum was also red-shifted with respect to the spectra of the individual constituents, indicating strong π–π stacking and thus good contact between the constituents. The Raman spectrum (Figure 3b) of the dried composite thin film also shows signatures of all three constituents. The peak at 1464 cm^{-1} is attributable to the stretching mode of C_{60} cages. Two dominant peaks at 1596 and 1587 cm^{-1} is due to phonon transition within the SWCNTs, and a D peak at 1350 cm^{-1} is due to alternating sp^2/sp^3 clusters of GO.[28] The Raman spectrum of the composite is also red-shifted with respect to the Raman spectra of the individual constituents, which confirms strong π–π stacking. Furthermore, the D band is absent from the Raman spectrum of SWCNT, which confirms that the SWCNT are unfunctionalized.

All-Carbon Photovoltaic Cell

Thin films of C_{60}/SWCNT/GO prepared by spin-coating have very low roughness (Figure 1f). Therefore, the film can be used as the active layer of photovoltaic devices. An exploded schematic of the layer structure of the photovoltaic is shown in Figure 4a. ITO/glass substrates pre-treated with a smooth 20 nm layer of PEDOT:PSS were used. To deposit the active layer, C_{60}/SWCNT/GO hybrid dispersion was spin-coated onto the substrate, forming a thin film of 5–8 nm. Thermal reduction at 150 °C converted GO to r-GO, and rendered the composite conductive. An additional 15 nm layer of C_{60} (e-C_{60}) was deposited atop the active layer via thermal evaporation. The additional C_{60} layer acts as a hole-blocking and electron-transporting layer. To complete the device, 100 nm of Al was deposited as the electrodes of the device.

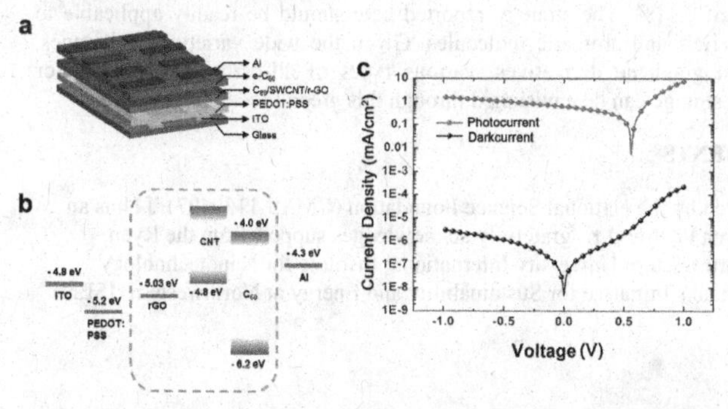

Figure 4. All-carbon photovoltaic device. (a) Schematic diagram illustrating the photovoltaic device structure with C_{60}/SWCNTs/r-GO as the active layer and an additional layer of thermally evaporated C_{60} as the protection/hole blocking layer. (b) Corresponding schematic energy level diagram of the device. The all-carbon composite is shown in the dashed box. The ionization potential of the SWCNTs sample is experimentally determined from a thin film comprised of mixture of semiconducting and metallic nanotubes. LUMO of SWCNTs is not marked because their band gap varies by diameter. (c) Representative I-V characteristics measured in the dark and under illumination.

The corresponding band diagram of the device is shown in Figure 4b. The work functions of ITO, PEDOT:PSS, and Al, and the energy level of the highest occupied molecular orbital (HOMO) and the lowest unoccupied molecular orbital (LUMO) of C_{60} are taken from literature.[14,29,30] The ionization potentials of r-GO and SWCNT were determined by ultraviolet photoelectron spectroscopy (UPS). We estimate the IP represents the average HOMO level of the semiconducting carbon nanotubes. The LUMO is not marked due to the polydispersity of carbon nanotube diameters, which results in different band gaps. The energy level difference at the C_{60}/SWCNT interface is sufficiently large to dissociate photogenerated excitons. Electrons propagate via the e-C_{60} layer towards the Al electrodes, while holes are transported via the SWCNT/r-GO network towards the ITO electrodes. Thus, a photocurrent is generated. The performance of the device is shown in Figure 4c. The current-voltage response shows distinct diode behavior, with a rectification ratio of 10^4 for the dark current. Under illumination, the device displays an on/off ratio of nearly 10^6, which surpasses the performance of devices that use covalently linked, polymer-wrapped, or surface functionalized all-carbon composites.[1,4-6] Moreover, the device shows a short-circuit current (J_{sc}) of 1.23 mA/cm^2, an open circuit voltage (V_{oc}) of 0.59 V, and fill factor of 0.29, resulting in a power conversion efficiency of 0.21%. This photovoltaic response is unprecedented for devices that use an all-carbon active layer.[1-7]

CONCLUSION

In conclusion, all-carbon composites made of C_{60}, SWCNTs and GO were created using a surfactant-free, aqueous processing approach. Mild thermal annealing renders the composite conductive, and the photoconductive all-carbon composite can be incorporated as the active layer of photovoltaic devices. The photovoltaic devices displayed an unprecedented photoconductive response with an on/off ratio of nearly 10^6, V_{oc} of 0.59 V, and power conversion efficiency of 0.21%. The strategy reported here should be readily applicable to other carbon-rich materials and aromatic molecules. Given the wide variety of fullerenes, carbon nanotubes, and graphene derivatives, various types of all-carbon composites for energy conversion and storage can be envisioned through this green processing route.

ACKNOWLEDGEMENTS

This work was supported by the National Science Foundation (CMMI 1130407). J.H. is an Alfred P. Sloan Research Fellow. J.K. gratefully acknowledges support from the Ryan Fellowship and the Northwestern University International Institute for Nanotechnology. A.C.L.T. and V.C.T. thanks Initiative for Sustainability and Energy at Northwestern (ISEN) for support.

REFERENCES

(1) Yamamoto, Y. Fukushima, T. Suna, Y. Ishii, N. Saeki, A. Seki, S. Tagawa, S. Taniguchi, M. Kawai, T.; Aida, T. *Science* **2006**, *314*, 1761 -1764.
(2) Nasibulin, A. G. Pikhitsa, P. V. Jiang, H. Brown, D. P. Krasheninnikov, A. V. Anisimov, A. S. Queipo, P. Moisala, A. Gonzalez, D. Lientschnig, G. Hassanien, A. Shandakov, S. D. Lolli, G. Resasco, D. E. Choi, M. Tomanek, D.; Kauppinen, E. I. *Nat Nano* **2007**, *2*, 156-161.
(3) Umeyama, T. Tezuka, N. Fujita, M. Hayashi, S. Kadota, N. Matano, Y.; Imahori, H. *Chem. Eur. J.* **2008**, *14*, 4875-4885.
(4) Kalita, G. Adhikari, S. Aryal, H. R. Umeno, M. Afre, R. Soga, T.; Sharon, M. *Appl. Phys. Lett.* **2008**, *92*, 063508.
(5) Yamamoto, Y. Zhang, G. Jin, W. Fukushima, T. Ishii, N. Saeki, A. Seki, S. Tagawa, S. Minari, T. Tsukagoshi, K.; Aida, T. *Proceedings of the National Academy of Sciences* **2009**, *106*, 21051-21056.
(6) Umeyama, T. Tezuka, N. Seki, S. Matano, Y. Nishi, M. Hirao, K. Lehtivuori, H. Tkachenko, N. V. Lemmetyinen, H. Nakao, Y. Sakaki, S.; Imahori, H. *Adv. Mater.* **2010**, *22*, 1767-1770.
(7) Zhang, X. Huang, Y. Wang, Y. Ma, Y. Liu, Z.; Chen, Y. *Carbon* **2009**, *47*, 334-337.
(8) Zhu, H. Wei, J. Wang, K.; Wu, D. *Solar Energy Materials and Solar Cells* **2009**, *93*, 1461-1470.
(9) Sariciftci, N. S. Smilowitz, L. Heeger, A. J.; Wudl, F. *Science* **1992**, *258*, 1474 -1476.
(10) Tans, S. J. Verschueren, A. R. M.; Dekker, C. *Nature* **1998**, *393*, 49-52.
(11) Dürkop, T. Getty, S. A. Cobas, E.; Fuhrer, M. S. *Nano Letters* **2004**, *4*, 35-39.
(12) Gilje, S. Han, S. Wang, M. Wang, K. L.; Kaner, R. B. *Nano Letters* **2007**, *7*, 3394-3398.
(13) Eda, G. Fanchini, G.; Chhowalla, M. *Nat Nano* **2008**, *3*, 270-274.
(14) Arnold, M. S. Zimmerman, J. D. Renshaw, C. K. Xu, X. Lunt, R. R. Austin, C. M.; Forrest, S. R. *Nano Letters* **2009**, *9*, 3354-3358.
(15) Kim, F. Cote, L. J.; Huang, J. *Advanced Materials* **2010**, *22*, 1954-1958.

(16) Kim, J. Cote, L. J. Kim, F. Yuan, W. Shull, K. R.; Huang, J. *Journal of the American Chemical Society* **2010**, *132*, 8180-8186.
(17) Cote, L. J. Kim, J. Tung, V. C. Luo, J. Kim, F.; Huang, J. *Pure Appl. Chem.* **2011**, *83*, 95-110.
(18) Schniepp, H. C. Li, J.-L. McAllister, M. J. Sai, H. Herrera-Alonso, M. Adamson, D. H. Prud'homme, R. K. Car, R. Saville, D. A.; Aksay, I. A. *The Journal of Physical Chemistry B* **2006**, *110*, 8535-8539.
(19) Stankovich, S. Dikin, D. A. Dommett, G. H. B. Kohlhaas, K. M. Zimney, E. J. Stach, E. A. Piner, R. D. Nguyen, S. T.; Ruoff, R. S. *Nature* **2006**, *442*, 282-286.
(20) Cote, L. J. Cruz-Silva, R.; Huang, J. *Journal of the American Chemical Society* **2009**, *131*, 11027-11032.
(21) Hummers, W. S.; Offeman, R. E. *Journal of the American Chemical Society* **1958**, *80*, 1339.
(22) Cote, L. J. Kim, F.; Huang, J. *Journal of the American Chemical Society* **2009**, *131*, 1043-1049.
(23) Kim, F. Luo, J. Cruz-Silva, R. Cote, L. J. Sohn, K.; Huang, J. *Adv. Funct. Mater.* **2010**, *20*, n/a-n/a.
(24) Erickson, K. Erni, R. Lee, Z. Alem, N. Gannett, W.; Zettl, A. *Adv. Mater.* **2010**, n/a-n/a.
(25) Hare, J. P. Kroto, H. W.; Taylor, R. *Chemical Physics Letters* **1991**, *177*, 394-398.
(26) Ausman, K. D. Piner, R. Lourie, O. Ruoff, R. S.; Korobov, M. *The Journal of Physical Chemistry B* **2000**, *104*, 8911-8915.
(27) Li, D. Muller, M. B. Gilje, S. Kaner, R. B.; Wallace, G. G. *Nat Nano* **2008**, *3*, 101-105.
(28) Dresselhaus, M. S. Jorio, A. Hofmann, M. Dresselhaus, G.; Saito, R. *Nano Letters* **2010**, *10*, 751-758.
(29) Sato, N. Saito, Y.; Shinohara, H. *Chemical Physics* **1992**, *162*, 433-438.
(30) Shirley, E. L.; Louie, S. G. *Phys. Rev. Lett.* **1993**, *71*, 133.

Mater. Res. Soc. Symp. Proc. Vol. 1344 © 2011 Materials Research Society
DOI: 10.1557/opl.2011.1357

1/*f* Noise in Graphene Field-Effect Transistors: Dependence on the Device Channel Area

Guanxiong Liu[1], Sergey Rumyantsev[2,3], William Stillman[2], Michael Shur[2] and Alexander A. Balandin[1]
[1]Department of Electrical Engineering, University of California, Riverside, California 92521
[2]Rensselaer Polytechnic Institute, Troy, New York 12180
[3]Ioffe Institute, Russian Academy of Sciences, St. Petersburg, 194021 Russia

ABSTRACT

We carried out a systematic experimental study of the low-frequency noise characteristics in a large number of single and bilayer graphene transistors. The prime purpose was to determine the dominant noise sources in these devices and the effect of aging on the current-voltage and noise characteristics. The analysis of the noise spectral density dependence on the surface area of the graphene channel indicates that the dominant contributions to the 1/*f* electronic noise come from the graphene channel region itself. Aging of graphene transistors due to exposure to ambient for over a month resulted in substantially increased noise, which was attributed to the decreasing mobility of graphene and increasing contact resistance. The noise spectral density in both single and bilayer graphene transistors shows a non-monotonic dependence on the gate bias. This observation confirms that the 1/*f* noise characteristics of graphene transistors are qualitatively different from those of conventional silicon metal-oxide-semiconductor field-effect transistors.

INTRODUCTION

The extremely high mobility and saturation velocity make graphene a promising material for the high-frequency and communication applications. A triple-mode graphene amplifier, which capitalizes on the ambipolar nature of graphene has already been demonstrated [1]. Most of the proposed communication applications of graphene require a low-level of the low-frequency noise. The low-frequency noise can be up-converted due to device non-linearity, and contributes to the phase noise of the systems. We have previously reported measurements of the noise in the bottom-gate [2] and top-gate graphene transistors [3]. At the same time, a number of issues such as the effects of surface, contacts, environmental exposure, and temperature on the noise level are still remain open. The exact mechanism of 1/f noise in graphene is also not known.

Here we describe the results of the measurements in a large number of graphene devices over an extended time period, which allowed us to elucidate the effects of environmental exposure. For all examined devices the noise spectra for the frequency range 1 Hz-50 kHz were close to the $1/f^{\gamma}$ with γ=1.0-1.1 [4]. Aging of graphene transistors for over a month resulted in substantially increased noise attributed to the decreasing mobility of graphene and increasing contact resistance. The noise spectral density in both single layer graphene (SLG) and bilayer graphene (BLG) transistors either increased with deviation from the charge neutrality point or depended weakly on the gate bias. This observation confirms that the low-frequency noise characteristics of graphene transistors are qualitatively different from those of conventional silicon devices.

EXPERIMENTAL DETAILS

Graphene samples were produced by the mechanical exfoliation from bulk highly oriented pyrolitic graphite (HOPG). The graphene flakes were placed on the standard Si/SiO2 substrates during exfoliation using the standard procedure and initially identified with an optical microscope. SLG and BLG sample were selected using micro-Raman spectroscopy through the 2D/G'-band deconvolution [5, 6]. Electron beam lithography (EBL) is used to define the source and drain areas through the contact bars with the help of pre-deposited alignment marks. The 8 nm Ti/ 80 nm Au metals were sequentially deposited on graphene by the electron-beam evaporation (EBE). In this design, the degenerately doped Si substrate acted as a back gate. The micro-Raman inspection was repeated after the fabrication and electrical measurements to make sure that graphene's crystal lattice was not damaged.

The current – voltage (I-V) characteristics were measured using a semiconductor parameter analyzer (Agilent 4156B). The characteristics were studied for the pristine (as soon as fabricated) and aged graphene transistors kept in ambient environment over a month time period. Figure 1 shows two of the devices we fabricated before and after aging. We notice that the current decreases substantially after the device aging. This degradation can be attributed to the increased contact resistance. We also notice that the Dirac point of graphene device either shift up or down after aging, which is probably the results of unintentional absorbates that attached on graphene flake during aging time.

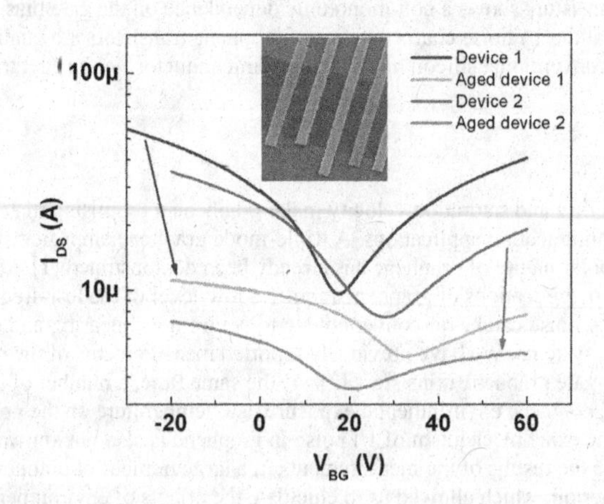

Figure 1: Transfer characteristics of two SLG graphene transistors before and after aging. Drain voltage V_{DS}=100mV. The inset shows a SEM image of graphene devices.

The low-frequency noise was measured before and after device aging in ambient condition. The frequency range we performed noise measurement is from 1 Hz to 50 kHz. The graphene transistors were biased in a common source mode at source-drain bias V_{DS} = 50 mV. The voltage-referred electrical current fluctuations S_V from the load resistor R_L connected in series with the drain were analyzed by a SR770 FFT spectrum analyzer. The noise spectrum of graphene device within our measurement range shows 1/f dependence on frequency, shown in Figure 2. A typical graphene device is shown in the inset of Figure 2.

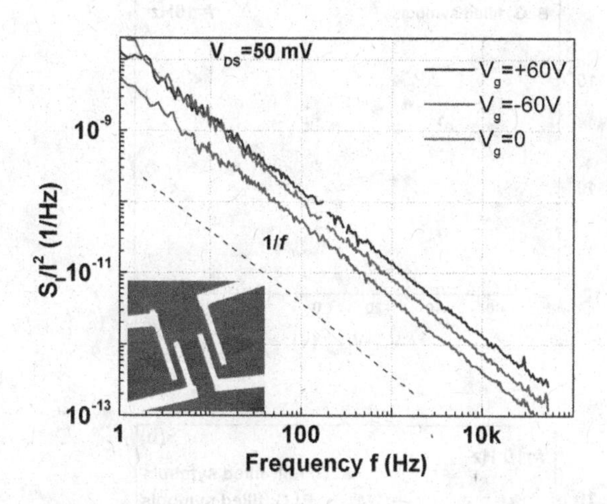

Figure 2: Noise spectral density of graphene device within the frequency range from 1 Hz to 100 kHz. The spectrum are close to the $1/f^{\gamma}$ with γ=1.0-1.1. The inset shows the optical image of a typical graphene device.

DISCUSSION

In conventional semiconductor MOSFETs, the gate voltage dependence of noise spectral density can provide useful information about the noise sources and mechanisms. The low-frequency noise is usually analyzed in the framework of the McWhorter model [7]. In this model, the low-frequency noise is caused by the capture and emission process of the carriers in the channel by the traps in the oxide. Therefore the trap concentration in the oxide is a natural figure-of-merit for the noise amplitude in MOSFETs. The McWhorter model predicts that the normalized noise spectral density, S_I/I_d^2, decreases in the strong inversion regime as $\sim 1/n_s^2$ (where n_s is the channel carriers concentration). Any deviation from this law might indicate the influence of the contacts, non-homogeneous trap distribution in energy or space, or contributions of the mobility fluctuations to the current noise [8-10].

We measured the gate dependence of noise of many SLG and BLG devices, shown in Figure 3(a). Note that we plot the gate voltage by (V_{BG}-V_{Dirac}), therefore each side away from zero point corresponds to increasing of carrier concentration. We notice that all of our devices have non-monotonic gate dependence and asymmetry between electron and hole sides of the current-voltage characteristics. These behaviors indicate a complicated noise behavior of graphene device, which is qualitatively different from conventional MOSFET in differentiate the noise contribution from channel and contacts.

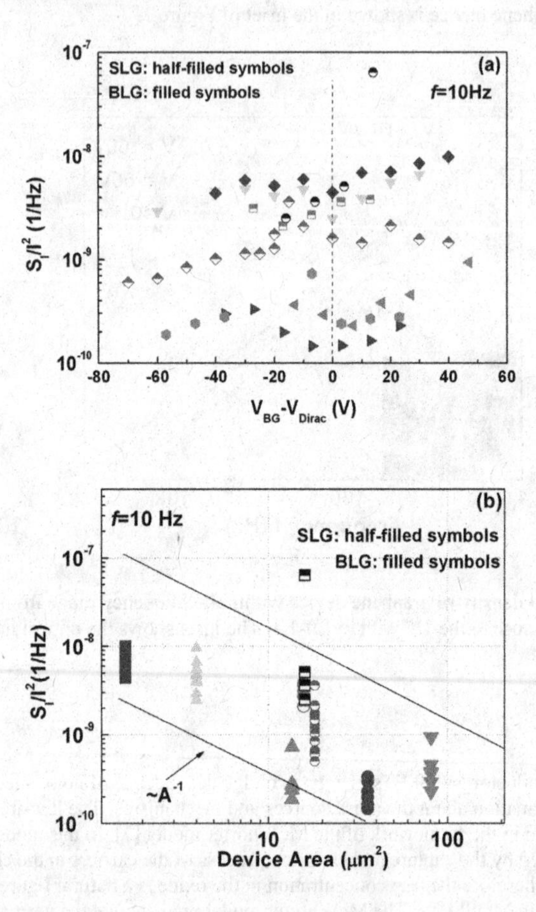

Figure 3: (a) Gate dependence of noise spectral density of SLG (half-filled symbols) and BLG (filled symbols) devices. Non-monotonic dependence on the gate bias suggests that the noise behavior in graphene device is qualitatively different from that in conventional MOSFETs. (b) The area dependence of the $1/f$ noise shows that the noise is inversely proportional to the channel size, which indicates that the noise in graphene device is mainly comes from graphene channel region itself rather than from the contacts.

As we measured noise from devices with a variety of channel size ranging from 1.5 to 80 μm², we notice that the noise has a relative large spread. However, interesting thing appears as we plot the area dependence of noise spectral density, S_I/I_d^2 for the same data from SLG and BLG devices shown in Figure 3(a). One can see from Figure 3(b) that the noise spectrum is roughly inversely proportional to device channel area. This area dependence suggests that the noise contribution should mainly come from graphene channel region. We know that graphene/SiO2 interface is not perfectly smooth, and the roughness creates electron-hole puddles [11], which act as scattering centers that fluctuates the carrier transport on graphene. A smoother substrate that is more compatible to graphene may reduce the noise in graphene-based device.

As mentioned before, aging of the graphene transistors due to the ambient exposure leads to the decrease of the carrier mobility and increase of the contact resistance. The noise measurements of the "aged" transistors have shown a substantial increase in the low-frequency noise. Figure 4 shows the gate dependence of the noise spectral density of the transistors before and after kept in ambient for about 1month. Since both the contact resistance and mobility degrade as a result of the environment exposure we assume that both contacts and graphene layer itself contribute the 1/f noise increase.

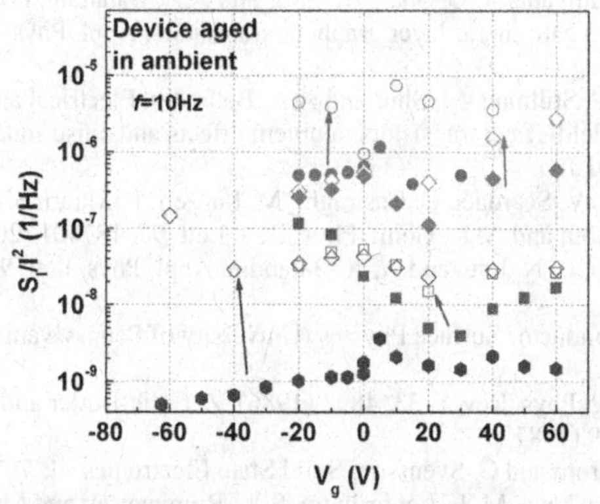

Figure 4: Substantial noise increases in graphene devices as a result of exposure to ambient. The filled symbols are before aging and the open symbols are after aging. The arrows indicate the transition of the same device.

CONCLUSIONS

We investigated low-frequency 1/*f* noise in graphene back-gate transistors focusing on the effects of environmental exposure and analysis of the noise sources. The exposure of the graphene transistors to ambient for a period of about one month resulted in substantial increase of the contact resistance, mobility degradation, as well as low-frequency noise. Through analysis of the noise spectral density dependence on the graphene channel area we find that the main contributions to noise come from graphene channel region. For all examined graphene transistors, both SLG and BLG, we found that noise has a non-monotonic dependence on the gate-bias. The observed noise behavior is very different from that in conventional Si MOSFETs.

ACKNOWLEDGMENTS

The work at UCR was supported, in part, by DARPA – SRC Focus Center Research Program (FCRP) through its Center on Functional Engineered Nano Architectonics (FENA) and DARPA Defense Microelectronics Activity (DMEA) under agreement number H94003-10-2-1003. The work at RPI was supported by the National Science Foundation under I/UCRC "Connection One" and by IFC. This work was also partially supported by RFBR (Grant No. 08-02-00010).

REFERENCES

1. X. Yang, G. Liu, A.A. Balandin and K. Mohanram, Triple-mode single-transistor graphene amplifier and its applications, ACS Nano, **4**, 5532 (2010).
2. Q. Shao, G. Liu, D. Teweldebrhan, A. A. Balandin, S. Rumyantsev, M. Shur and D. Yan, Flicker noise in bilayer graphene transistors, Electron Device Lett., **30**, 288 (2009).
3. G. Liu, W. Stillman, S. Rumyantsev, Q. Shao, M. Shur and A.A. Balandin, Low-frequency electronic noise in the double-gate single-layer graphene transistors, Appl. Phys. Lett., **95**, 033103 (2009).
4. S. Rumyantsev, G. Liu, W. Stillman, M. Shur and A.A. Balandin, Electrical and noise characteristics of graphene field-effect transistors: ambient effects and noise sources, J. Physics: Cond. Matt., **22**, 395302 (2010).
5. A. C. Ferrari, J. C. Meyer, V. Scardaci, C. Casiraghi, M. Lazzeri, F. Mauri, S. Piscanec, D. Jiang, K. S. Novoselov, S. Roth and A. K. Geim, Phys. Rev. Lett. **97**, 187401 (2006).
6. I. Calizo, F. Miao, W. Bao, C. N. Lau, and A. A. Balandin, Appl. Phys. Lett. **91**, 071913 (2007).
7. A. L. McWhorter, Semiconductor Surface Physics (University of Pennsylvania Press, Philadelphia, 1957), pp 207.
8. C. Surya and T. Y. Hsiang, Phys. Rev. B 33, 4898 (1986); Z. Celik-Butler and T.Y. Hsiang, Solid State Electron., **30**, 419 (1987).
9. S. Christensson, I. Lundstrom and C. Svensson, Solid State Electronics 11, 797 (1968).
10. A.P. Dmitriev, E. Borovitskaya, M. E. Levinshtein, S. L. Rumyantsev and M. S. Shur, J. Appl. Phys. **90**, 301 (2001).
11. Y. Zhang, V. W. Brar, C. Girit, A. Zettl and M. F. Crommie, "Origin of spatial charge inhomogeneity in graphene", Nat. Phys. **5**, 722 (2009).

Mater. Res. Soc. Symp. Proc. Vol. 1344 © 2011 Materials Research Society
DOI: 10.1557/opl.2011.1352

Numerical Study of Scaling Issues in Graphene Nanoribbon Transistors

Man-Tieh Chen and Yuh-Renn Wu *
Graduate Institute of Photonics and Optoelectronics and Department of Electrical Engineering, National Taiwan University, Taipei, Taiwan, 10617.
* yrwu@cc.ee.ntu.edu.tw

ABSTRACT

This paper addresses scaling issues in graphene nanoribbon transistors (GNRFETs) by using a two-dimensional (2-D) Poisson and drift-diffusion solver with finite element method (FEM). GNRFETs with the back gate control and the channel width down to less than 5nm have been reported to have I_{on}/I_{off} ratio up to 10^6. Our simulations show an agreement with the published experimental work and show a potential to reach unit current gain cut-off frequency, f_T, up to more than 1THz with a satisfying I_{on}/I_{off} ratio at the same time. This makes GNRFETs attractive for high speed logic.

INTRODUCTION

In high speed RF applications, the lack for terahertz source has been a long term challenge for people working on this field. The feasibility of making graphene transistor has opened the possible way to overcome the terahertz barrier. Graphene is a single layer graphite and features many exceptional properties[1–3]. Due to the linear E-k relation and almost zero bandgap, the mobility of graphene is quite high, which has been experimentally demonstrated the cut-off frequency f_T over 230GHz[4]. The potential f_T could be even higher with a proper design. However, due to the zero bandgap, the I_{on}/I_{off} ratio is bad and limits its potential application. There are several approaches to open a bandgap : (1) applying perpendicular electric field to bilayer graphene[5, 6]; (2) forming an epitaxial graphene on silicon carbide[7, 8]; (3) water adsorption to the graphene surface[9]; (4) applying strain to graphene[10]; and (5) patterned hydrogen adsorption in graphene[11]. Theory and experiment have indicated that if we use the vertical electric field, bandgap values can be opened up to 0.2eV. However, the approaches are not feasible for a circuit design since the high electric field and water are not welcomed for most devices. Theory predicted that to open a bandgap by applying strain requires a global uniaxial strain exceeding 20%, which is difficult to achieve in practice. Also, except for the patterned hydrogen adsorption in graphene, these approaches mentioned above cannot provide an adequate bandgap for logic applications.

Theoretical study[12] has indicated that the Graphene nanoribbon (GNR) with channel width below 10nm can open up a bandgap and become semiconductor type material. Back-gated transistors based on GNR have been experimental reported to have I_{on}/I_{off} ratio up to 10^6[13]. Recently, to obtain larger I_{on}/I_{off} and smaller subthreshold swing with a small bandgap, tunnel FETs have been studied in simulations[14, 15] but have not been experimentally reported yet. In this study, we are interested in investigating the potential high speed performance of the GNR transistor.

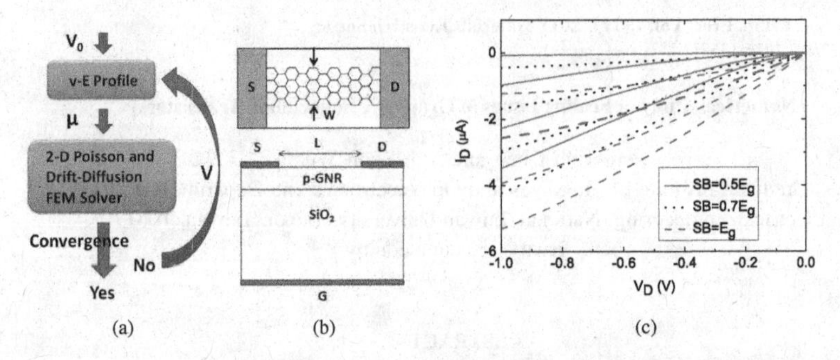

FIG. 1: (a) The schematic flowchart of the self-consistent iteration loop and simulation process of our built-models. (b) Schematic cross section of the GNRFET for simulation in this work. (c) The I_D-V_D characteristic for GNRFETs with various Schottky barrier height of both source and drain contact. The nanoribbon width is 2.2nm and the effective bandgap is 0.497eV.

FORMALISM

We use the 2-D finite element (FEM) Poisson and drift-diffusion solver to obtain the I-V curves, charge distribution, and electric field along the channel. Field-dependent mobility models derived from Monte-Carlo simulation results[16–20] are employed to the 2-D Poisson and drift-diffusion solver[21, 22] and the flowchart is shown in Fig.1 (a). By using the v-E profiles as the input mobility model, the coupled Poisson and drift-diffusion equations are solved in each self-consistent iteration loop. In the lack of a proper experimental study, we set the saturation velocity to be 4.4×10^7cm/s from Monte-Carlo results of graphene[16, 17]. The low-field mobility μ is derived from Monte-Carlo results of graphene nanoribbons[18–20] by taking the following mechanisms into account: (1) acoustic phonon scattering; (2) polar optical phonon scattering; and (3) line edge roughness scattering.

Our simulated device is a back-gated transistor with a p-type channel. The GNR with different widths from 1.1nm to 4nm are used. An 8nm thick SiO_2 is used as the gate insulator, as shown in Fig. 1(b). In the simulation, the drain and source contact of the GNR transistor is Schottky without intentionally doping in the contact region. We applied the nearest-neighbor tight binding method to extract dispersion curves of an armchair GNR, where the quantized confined wavevector is used. Therefore the subbands of the dispersion curves for the armchair GNR with different channel widths (represented as N dimer lines width[12]) can be obtained, and the effective mass and the bandgap of semiconducting GNRs with different widths can be estimated, and is shown in Table I. The bandgaps from first principles calculations[12] are also listed for comparison.

RESULTS AND DISCUSSION

Due to the difficulties of doping and the small bandgap nature of graphene nanoribbon. The contact of drain and source might be either Schottky contact or ohmic contact depending on the

TABLE I: The calculated effective mass and bandgap from tight binding method and first principles for GNRs with different channel widths.

dimer lines	W (nm)	E_g (eV)	$m_e^* = m_h^*$ (m_0)	*ab initio* E_g[12] (eV)	Mobility (cm²/Vs
N_a=10	1.1	0.914	0.127	1.05	
N_a=19	2.2	0.497	0.063	0.55	800 [20]
N_a=34	4.0	0.282	0.034	0.30	4200[20]

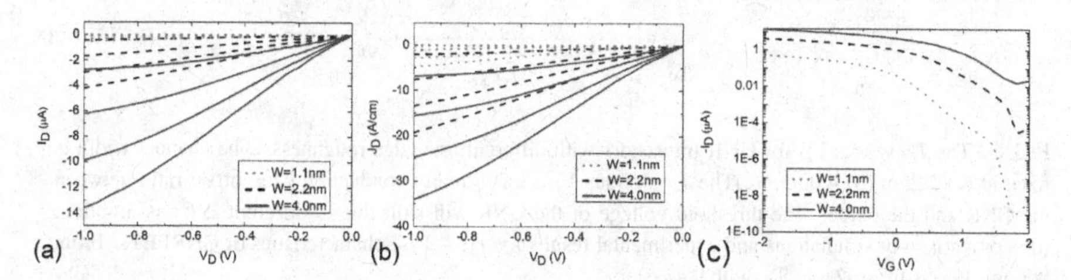

FIG. 2: (a) The drain current I_D versus V_D curves. (b) The drain current density J_D versus V_D curves. (c) The I_D versus V_G curves for the GNR transistors with different channel width.

metal work function. Therefore, it is important to understand the influence of the Schottky contact behavior at the drain and source. Figure. 1(c) shows the I_D-V_{DS} with different source/drain Schottky barrier height to the conduction band. Because the device is operated as a PMOS, the total drain current decreases when the smaller work function metal is used. Besides, for the smaller Schottky barrier height to the conduction band, the electron leakage current is observed with larger positive bias.

Figures 2 (a) and 2 (b) show the I_D-V_{DS} curves and J_D-V_{DS} curves with different channel widths, respectively. Since the wider channel width has smaller effective mass and higher mobility, the current density is larger, which is also confirmed by[18]. However, the device is hard to pinch off for the wider channel width and the I_{on}/I_{off} ratio is low, as shown in Fig. 2 (c). The wider channel with smaller bandgap forms smaller Schottky barrier at source and drain, resulting in larger thermal emission electron current and larger off-state current density. When the channel width becomes smaller, the effective mass becomes larger and the bandgap increases as well. The mobility of GNR becomes smaller compared to that of graphene due to larger effective mass and edge roughness scattering. However, we get a better I_{on}/I_{off} ratio and a better transistor property.

When the device channel width is shrinking, the influence of edge effect will become very significant. Fig. 3 (a) shows the I_D-V_{DS} curves for channel width, W, equal to 2.2nm under different edge roughness. The line edge roughness (LER) scattering model follows the studies in ref. [19, 20]. The result shows that the current density decreases to be less than 1/2 if the edge roughness is very large. In addition, the I_D-V_{DS} curves become more saturated when the edge roughness is smaller. Fig. 3 (b) shows the I_D-V_{DS} characteristic for channel width W equal to 4.0nm under different edge roughness. It can be observed that the line edge roughness scattering becomes smaller in wider GNR, which is consistent to the expression of LER scattering rate in [19].

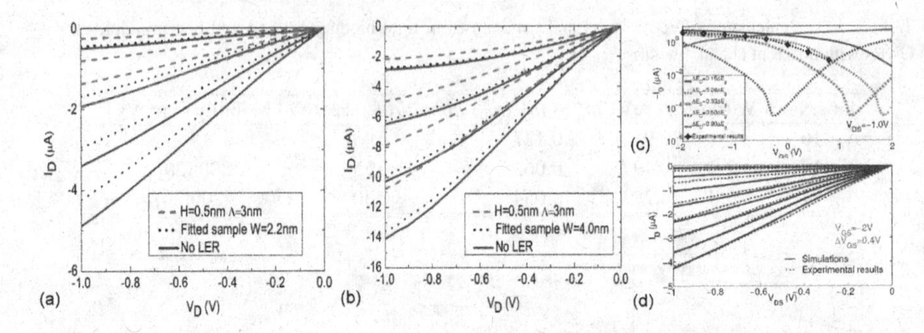

FIG. 3: The I_D versus V_D for GNR transistors with different line edge roughness. The channel width is fixed at (a) 2.2nm (b) 4.0nm. (c)The I_{DS} versus V_{DS} for different conduction band offset ratio between the GNR and the oxide. The threshold voltage of the GNR will shift due to different ΔE_c assumption. (d) Comparison of simulations and experimental results for $I_D - V_{DS}$ characteristics of GNRFETs. From bottom, V_{GS} is from -2 to 0.8V, with 0.4V/step.

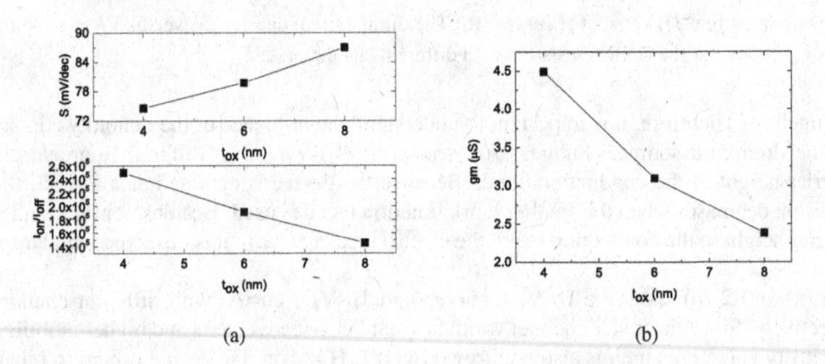

FIG. 4: (a) The subthreshold swing S, I_{on}/I_{off} ratio and (b) The transconductance g_m versus the gate oxide thickness. The gate oxide dielectric constant is 3.9. Drain voltage V_d=-0.6V.

To simulate the I-V characteristics, it is important to know the conduction band offset ratio between the SiO_2 and GNR. To obtain this value, we tried different values of band offset and compared to the experimental work from[13], as shown in Fig. 3(c) . The results show that the conduction band offset ratio ($\Delta E_c/\Delta E_g$) equal to 0.26 has the better fit in the $I_D - V_{DS}$ characteristics, as shown in Fig. 3(d), where the fitted mobility is $800cm^2/Vs$ and effective mass is $0.063m_0$ for channel width equal to ~2.2nm.

After defining the device structure, we will investigate the potential RF performance of GNRFETs. A self-aligned FET has an advantage of suppressing the short channel effect and the parasitic resistance and parasitic capacitance between the gate and the source/drain. The channel width is set to be 2.2nm. The edge roughness of GNR is considered since it is usually hard to avoid. When the mean free path of the carrier is much longer than the channel length, the velocity

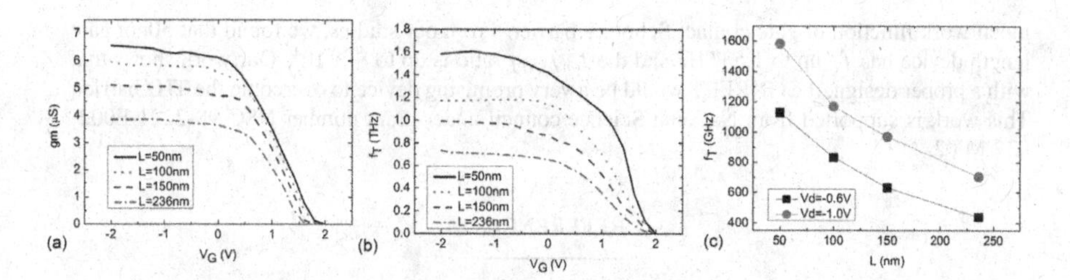

FIG. 5: (a) The transconductance versus gate bias at V_{DS}= -1V for GNRFETs with different gate length. (b) The cut off frequency versus gate bias at V_{DS}= -1V for GNRFETs with different gate length. (c) Calculated maximum cut off frequency for GNRFETs versus different gate lengths.

overshoot should be considered.

Fig. 4(a) and Fig. 4(b) shows the subthreshold swing S, I_{on}/I_{off} ratio, and transconductance g_m versus gate oxide thickness. We observe that the decrease of gate oxide thickness leads to a better gate control, smaller subthreshold swing and larger I_{on}/I_{off} ratio. For t_{OX}=4nm, the subthreshold swing is about 74mV/decade and the I_{on}/I_{off} ratio is up to 2.5×10^5 (same measurement range as [13]). The transconductance g_m increases as the t_{OX} decreases. The g_m is up to 4.5μS for t_{OX}=4nm.

We investigated the scaling characteristics of the GNRFET channel length. The unity current gain cut off frequency, f_T, and transconductance, g_m, as a function of the gate voltage with various channel length L are shown in Figs. 5 (a) and 5(b). The transconductance increases as the channel length is shortened. The cut off frequency is calculated by $f_T = v_{ave}/2\pi L_{DS}$, and carrier drift velocity is obtained from simulation results. The down scaling trend of the f_T is depicted in Fig. 5(c). For V_d equal to -0.6V, the f_T of graphene nanoribbon transistor with $L = L_{DS}$ equal to 236nm is 438GHz. For V_D equal to -1.0V, the f_T with the same channel length is 700GHz, compared to the fast reported graphene transistor so far with f_T=230GHz with L equal to 240nm[4]. The reason for the larger predicted f_T compared to experimental work may be due to the smaller contact parasitic resistance, where the contact resistance will reduce the electric field induced in the channel.

Our studies also show that with 50nm gate length for V_d=-0.6V, the f_T is up to 1.1THz, where the I_{on}/I_{off} ratio is 2.5×10^5. With the same gate length for V_d=-1.0V, the estimated f_T is as high as 1.58 THz, where the I_{on}/I_{off} ratio still maintains about 8×10^4. However, this result is based on that the contact is a perfect ohmic behavior. And there is no reverse tunneling leakage of electron current to cancel the hole current gain and transconductance value.

CONCLUSIONS

In this paper, numerical studies of graphene nanoribbon FETs have been presented. By using tight-binding method, effective mass and bandgap values of GNRs with different widths are extracted for the simulation. We adopted various field-dependent mobility models to study the GNR transistor's properties in different source/drain Schottky barrier height, channel widths and line edge roughness. The fitted conduction band offset ratio is 0.26. However, it depends on the actual metal work function of gate contact Schottky barrier. From our studies, we found that 50nm gate length device has f_T up to 1.58THz and the I_{on}/I_{off} ratio is up to 8×10^4. Our work shows that with a proper design, the GNRFET would be a very promising device to overcome the THz barrier. This work is supported from National Science council under grant number NSC-98-2221-E002-037-MY2.

REFERENCES

[1] Novoselov, K. S., Geim, A. K., Morozov, S. V., Jiang, D., Zhang, Y., Dubonos, S. V., Grigorieva, I. V., and Firsov, A. A. *Science* **306**(5696), 666–669 (2004).
[2] Geim, A. K. and Novoselov, K. S. *Nat Mater* **6**(3), 183–191 March (2007).
[3] Castro Neto, A. H., Guinea, F., Peres, N. M. R., Novoselov, K. S., and Geim, A. K. *Rev. Mod. Phys.* **81**(1), 109–162 Jan (2009).
[4] Avouris, P., Lin, Y., Xia, F., Farmer, D., Mueller, T., Dimitrakopoulos, C., and Jenkins, K. In *2010 IEEE International Electron Devices Meeting*, (2010).
[5] Castro, E. V., Novoselov, K. S., Morozov, S. V., Peres, N. M. R., dos Santos, J. M. B. L., Nilsson, J., Guinea, F., Geim, A. K., and Neto, A. H. C. *Phys. Rev. Lett.* **99**(21), 216802 Nov (2007).
[6] Gava, P., Lazzeri, M., Saitta, A. M., and Mauri, F. *Phys. Rev. B* **79**(16), 165431 Apr (2009).
[7] Zhou, S., Siegel, D., Fedorov, A., Gabaly, F., Schmid, A., Neto, A. C., Lee, D.-H., and Lanzara, A. *Nat Mater* **7**(4), 259–260 April (2008).
[8] Kim, S., Ihm, J., Choi, H. J., and Son, Y.-W. *Phys. Rev. Lett.* **100**(17), 176802 Apr (2008).
[9] Yavari, F., Kritzinger, C., Gaire, C., Song, L., Gulapalli, H., Borca-Tasciuc, T., Ajayan, P. M., and Koratkar, N. *Small* **6**(22), 2535–2538 (2010).
[10] Pereira, V. M., Castro Neto, A. H., and Peres, N. M. R. *Phys. Rev. B* **80**(4), 045401 Jul (2009).
[11] Balog, R., Jorgensen, B., Nilsson, L., Andersen, M., Rienks, E., Bianchi, M., Fanetti, M., Laegsgaard, E., Baraldi, A., Lizzit, S., Sljivancanin, Z., Besenbacher, F., Hammer, B., Pedersen, T. G., Hofmann, P., and Hornekaer, L. *Nat Mater* **9**(4), 315–319 April (2010).
[12] Son, Y.-W., Cohen, M. L., and Louie, S. G. *Phys. Rev. Lett.* **97**(21), 216803 Nov (2006).
[13] Wang, X. R., Ouyang, Y. J., Li, X. L., Wang, H. L., Guo, J., and Dai, H. J. *Physical Review Letters* **100**(20), 206803 May (2008).
[14] Lam, K.-T., Seah, D., Chin, S.-K., Bala Kumar, S., Samudra, G., Yeo, Y.-C., and Liang, G. *Electron Device Letters, IEEE* **31**(6), 555 –557 (2010).
[15] Zhang, Q., Fang, T., Xing, H., Seabaugh, A., and Jena, D. *Electron Device Letters, IEEE* **29**(12), 1344 –1346 (2008).
[16] Shishir, R. S. and Ferry, D. K. *Journal of Physics-condensed Matter* **21**(34), 344201 August (2009).
[17] Chauhan, J. and Guo, J. *Applied Physics Letters* **95**(2), 023120 July (2009).
[18] Bresciani, M., Palestri, P., Esseni, D., and Selmi, L. *Solid-State Electronics* **54**(9), 1015 – 1021 (2010).
[19] Fang, T., Konar, A., Xing, H., and Jena, D. *Phys. Rev. B* **78**(20), 205403 Nov (2008).
[20] Zeng, L., Liu, X., Du, G., Kang, J., and Han, R. In *International Conference on Simulation of Semiconductor Processes and Devices*, (2009).
[21] Wu, Y.-R., Singh, M., and Singh, J. *IEEE Trans. Electron Devices* **53**, 588–593 (2006).
[22] Wu, Y.-R. and Singh, J. *Journal of Applied Physics* **101**(11), 113712 (2007).

Mater. Res. Soc. Symp. Proc. Vol. 1344 © 2011 Materials Research Society
DOI: 10.1557/opl.2011.1073

Ultracapacitors Based on Graphene/MWNT Composite Films

Wei Wang[1], Shirui Guo[2], Jiebin Zhong[3], Jian Lin[3], Mihrimah Ozkan[1, 4], Cengiz Ozkan[1, 3]
[1] Materials Science & Engineering, University of California, Riverside, Riverside, CA, U. S. A.
[2] Chemistry, University of California, Riverside, CA, U. S. A
[3] Mechanical Engineering, University of California, Riverside, CA, U. S. A
[4] Electrical Engineering, University of California, Riverside, CA, U. S. A

ABSTRACT

Ultracapacitors are promising candidate for alternative energy storage applications since they can store and deliver energy at relatively high rates. In this work, we integrated large area CVD graphene with multi-walled carbon nanotubes (MWNTs) to fabricate highly conductive, large surface-area composite thin films used as electrodes in ultracapacitors. Uniform, large area graphene layers were produced by CVD on copper foils and were chemically modified. Chemically shortened MWNTs, ranging in length of 200~500 nm, were deposited by dropping on graphene layers. Graphene/MWNT composite films with different thicknesses were obtained. The surface morphology was investigated by SEM. The results demonstrated relatively dense and homogeneous net nanostructure. The measurements of cyclic voltammetry, chronopotentiometry, and electrochemical impedance spectroscopy (EIS) are conducted to determine its performance of graphene/MWNT film structures.

INTRODUCTION

Ultracapacitor is a promising alternative energy storage system due to its relatively fast rate of energy storage and delivery. The high energy and power density make the ultracapacitors suitable for a wide variety of applications [1]. The excellent chemical and physical properties of carbon materials such as high conductivity and surface-area, good corrosion resistance, excellent temperature stability, and relatively low cost make it a valuable candidate for the electrode of ultracapacitor [2]. Recently, intensive research has been conducted on carbon based supercapacitors, especially 1-D Carbon Nanotube (CNT) and 2-D graphene [3, 4]. The CNT-graphene hybrid structure is also a well studied area especially the patterned growth of vertically aligned CNT on graphene by chemical vapor deposition and also the solution processing of graphene-CNT hybrid materials [5, 6]. However, connecting CNT with CVD graphene without destroying their own advantages, i.e. high conductivity for graphene and large surface area for CNT is still a challenge. In this work, we reported a novel way to connect and stack chemically shortened multi-walled carbon nanotubes (MWNTs) to single layer CVD graphene (SLG) while keeping their own advantage for ultracapacitors. Electrochemical measurements show high energy density and specific capacitance of the as obtained SLG/MWNTs hybrid composite structure which demonstrates its promise as the electrode of ultracapacitors.

EXPERIMENT

The preparation of the electrode

Graphene films were grown on a 25 μm thick copper foil (Alfa Aesar, item No. 13382) by CVD method using a mixture of methane and hydrogen which originally presented by Xuesong Li, et al [7]. After the growth of graphene and cooling the system to room temperature, the graphene films were surface functionalized with 1-Pyrenebutyric Acid by dropping and spreading and then dried in a vacuumed chamber under room temperature for 4 hours. After surface functionalization the graphene films were rinsed and washed several times with methanol. The multi-wall carbon nanotubes (MWNTs) were chemically shortened by regular acid treatment and then after dilution and filtration the as obtained chemically shortened MWNTs were dispersed in DI water (1 mg/ml). MWNTs were coated on the graphene surface by dropping and the thickness of the CNT layer can be controlled by adjusting the concentration and amount of the MWNT solution.

The preparation and measurement of ultracapacitor Cells

The as prepared graphene/MWNTs composite films with copper substrates were devided into equal areas and attached on glass slides. The copper layer was clipped with a alligator clip and connected to a electrochemical analyzer (Gamry Reference 600™). Both the two glass slides and a piece of regular filter paper (Whatman 8-um filter paper) were soaked with the electrolyte (2 M Li_2SO_4 solution).

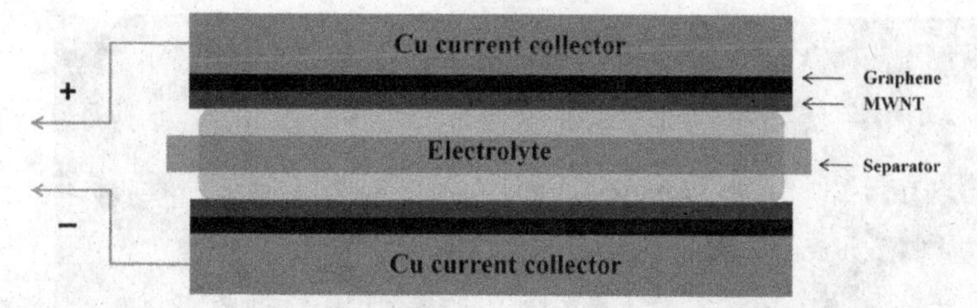

Figure 1. Schematic illustration of ultracapacitor cell structure based on single-layer graphene (SLG)/MWNT composite films.

A two-electrode measurement was employed for the electrochemical measurement, where the pretreated two glass slides were assembled into a sandwich structure with the as soaked filter paper works as a separator in between (as shown in Fig. 1). The cell was packaged with a parafilm and all the preparation and measurements were finished in nitrogen-filled glove box under room temperature. The measurements of cyclic voltammetry (CV), chronopotentiometry [charge-discharge (CD)], and electrochemical impedance spectroscopy (EIS) were conducted to show the performance of the ultracapacitor. The cyclic voltammetry scans were performed between -1 and 1 V at scan rates of 10, 20, 50, and 100 mV/sec. The charge-discharge curve was obtained at a current density of 1g/A. The potentiostatic EIS measurements were finished between 0.1 Hz and 1 MHz with amplitude of 10 mV.

DISCUSSION

Fig. 2.a, b shows a uniform and continuous surface morphology of the as grown single layer graphene on copper substrate which were used as the current collector of the ultracapacitor cell. By controlling the cooling rate the size of the copper grain can be controlled around 100 micrometers which is verified by Fig. 1a. The chemically shortened MWNT (1 mg/mL) solution was well dispersed and stable without precipitation even after 7 days. Fig.3 shows the acid-treated MWNTs dropped on SiO_2 substrate. It is clear that the diameter of the MWNTs (20~50 nm) are remained unchanged after treatment while the length was obviously shortened from 1.5 micrometer to 300~500 nm. Fig. d shows densely stacked and randomly oriented chemically shortened MWNTs on the surface of single layer graphene. The dense and homogeneous CNT net nanostructure produces more active sites as well as larger surface area which enable the material a good candidate for ultracapacitor electrode.

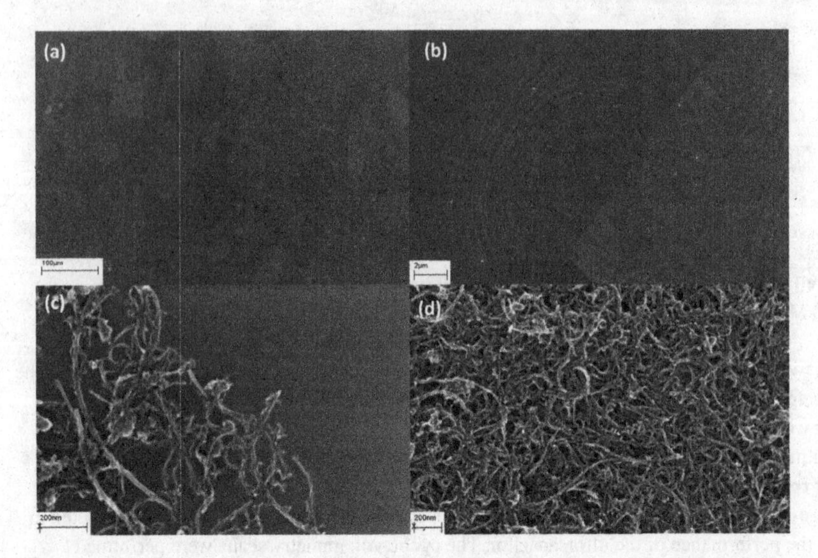

Figure 2 . SEM images of (a), (b) as grown single layer graphene on copper substrate, (c) chemically shortened multiwall carbon nanotubes on SiO_2 substratre, (d) chemically shortened MWNTs coated on graphene.

The cyclic voltammetric measurements of SLG/MWNTs composite films were performed from -0.8 V to 0.8 V at different sweep rates of 20, 50, 100, 200 mV/sec in 2 M Li_2SO_4 aqueous solution. Fig. 3 presents the CV curves, it can be seen that all of them show identical behaviors. The specific capacitance can be calculated from the curve by using the Equation 1.

$$C_s = \frac{\int IdV}{m \times \Delta V \times S} \tag{1}$$

Where C_s is the specific capacitance, $\int IdV$ is the integrated area of the CV curve, m is the mass of the active materials, ΔV is the potential range, and S is the scan rate. The specific capacitance of 134.38 F/g, 120.625 F/g, 119.38 F/g and 127.81 F/g were obtained for the sweep rates of 20, 50, 100, 200 mV/sec, respectively. This shows that the specific capacitances under different scan rates are approximately identical around the average specific capacitance of 125.55 F/g which suggest the as prepared ultracapacitor cell has fairly good stability and performance [8, 9].

The energy density was calculated by using Equation 2.

$$E = \frac{1}{2} C_s (\Delta V)^2 \qquad (2)$$

The energy density of 47.78 Wh/kg shows a good performance of the SLG/MWNTs ultracapcitor [4].

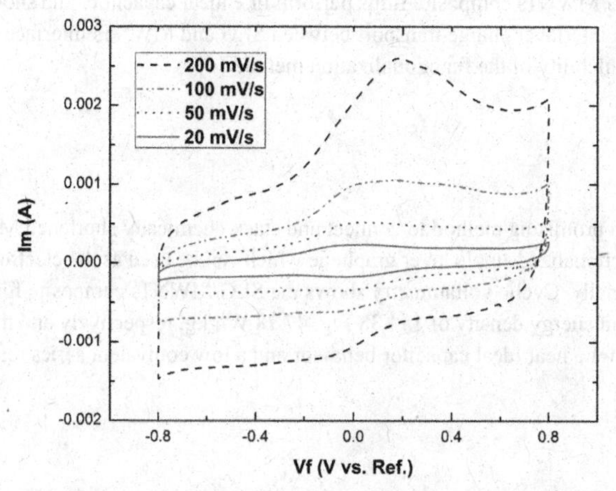

Figure 3. Cyclic voltammetric characteristics of single layer grpahene (SLG)/MWNTs composite thin films were measured from -0.8 V to 0.8 V at different sweep rates of 20, 50, 100, 200 mV/sec. 2 M Li_2SO_4 aqueous electrolyte.

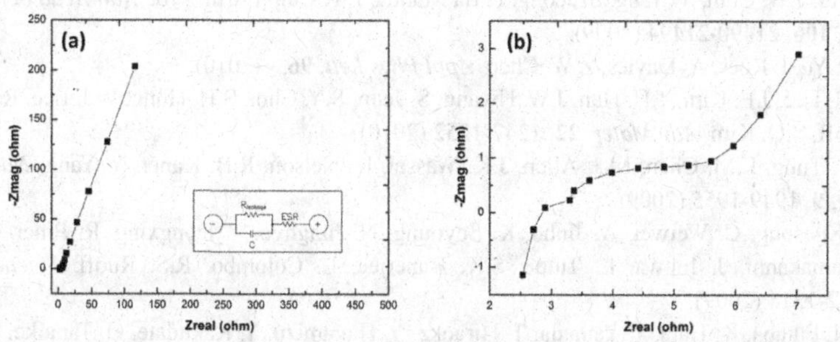

Figure 4. (a) The plot of the potentiostatic electrochemical impedance spectroscopy (EIS) which was performed between 0.1 Hz and 10^6 Hz with an amplitude of 10 mV. (b) shows the high frequency, low impedence region.

The potentiostatic electrochemical impedance spectroscopy (EIS) measurements were performed between 0.1 Hz and 1 MHz with amplitude of 10 mV. Fig. 4 a, b shows the potentiostatic EIS plot of the SLG/MWNTs composite film and the inset shows the simplified Randle's model which is the most common model fitted to ultracapacitor EIS spectra. Where *C* is the ideal capacitance, *ESR* is the equivalent series resistance, and $R_{leakage}$ is the leakage resistance. Generally, the near vertical impedance plots material will perform as an ideal capacitor which further present the SLG/MWNTs composite films perform like ideal capacitors and show a low ESR obtained from the interlayer charge transport between SLG and MWNTs interface which further confirm the availability of the functionalization method [4].

CONCLUSIONS

This work shows a promising method to connect and stack chemically shortened MWNTs directly to surface functionalized single layer graphene which can be used as the electrode of ultracapacitor. Additionally, Cyclic voltammetry shows the SLG/MWNTs composite films has a specific capacitance and energy density of 134.38 F/g, 47.78 Wh/kg, respectively and the potentialstatic EIS shows a near ideal capacitor behavior and a low equivalent series resistance.

REFERENCES

1. P. Simon, Y. Gogotsi, *Nat Mater*, **7**, 845-854 (2008).
2. A.G. Pandolfo, A.F. Hollenkamp, *J Power Sources*, **157**, 11-27 (2006).
3. L. Hu, J.W. Choi, Y. Yang, S. Jeong, F. La Mantia, L.F. Cui, Y. Cui, *Proc Natl Acad Sci U S A*, **106**, 21490-21494 (2009).
4. A.P. Yu, I. Roes, A. Davies, Z.W. Chen, *Appl Phys Lett*, **96**, - (2010).
5. D.H. Lee, J.E. Kim, T.H. Han, J.W. Hwang, S. Jeon, S.Y. Choi, S.H. Hong, W.J. Lee, R.S. Ruoff, S.O. Kim, *Adv Mater*, **22**, 1247-1252 (2010).
6. V.C. Tung, L.M. Chen, M.J. Allen, J.K. Wassei, K. Nelson, R.B. Kaner, Y. Yang, *Nano Lett*, **9**, 1949-1955 (2009).
7. L. Xuesong, C. Weiwei, A. Jinho, K. Seyoung, N. Junghyo, Y. Dongxing, R. Piner, A. Velamakanni, J. Inhwa, E. Tutuc, S.K. Banerjee, L. Colombo, R.S. Ruoff, *Science*, 1312-1314 (2009).
8. D.N. Futaba, K. Hata, T. Yamada, T. Hiraoka, Y. Hayamizu, Y. Kakudate, O. Tanaike, H. Hatori, M. Yumura, S. Iijima, *Nat Mater*, **5**, 987-994 (2006).
9. D. Yu, L. Dai, *The Journal of Physical Chemistry Letters*, **1**, 467-470 (2009).

Mater. Res. Soc. Symp. Proc. Vol. 1344 © 2011 Materials Research Society
DOI: 10.1557/opl.2011.1367

Graphene Oxide as a Two-dimensional Surfactant

Andrew R. Koltonow, Jaemyung Kim, Laura J. Cote, Jiayan Luo and Jiaxing Huang
Department of Materials Science and Engineering, Northwestern University, Evanston, USA

Abstract

Graphene oxide (GO) is a nonstoichiometric two-dimensional material obtained from the chemical oxidation and exfoliation of graphite, which has recently attracted intense research interest as a precursor for bulk production of graphene. GO has long been believed to be hydrophilic due to its dispersibility in water. Recent work in our group, however, has found that GO is actually a two-dimensional amphiphile; the edge of the sheet-like material is hydrophilic, while the basal plane of the material contains more hydrophobic graphitic nanodomains. To prove the concept, we demonstrate GO's surface activity at an air-water interface, as well as its utility in dispersing insoluble aromatic materials such as toluene, graphite, and carbon nanotubes in water. As a colloidal surfactant which can be converted to a conducting material, GO presents unique possibilities for aqueous solution processing of organic electronic materials.

Introduction

Graphene oxide (GO) is a nonstoichiometric two-dimensional carbon material which has recently attracted intense research interest as a precursor for bulk production of graphene. Typically GO is synthesized by treating graphite powders to strong oxidizing agents such as $KMnO_4$ in concentrated sulfuric acid.[1,2] This oxidizing treatment exfoliates the [002] planes of graphite atoms into water-dispersable single layers derivatized by carboxylic acids at the edges and phenol hydroxyl, and epoxide moieties mainly at the basal plane.[3,4] While extensive functionalization of the conjugated network renders GO insulating, conductivity can be partially restored through reduction by chemical,[5,6] thermal,[7,8] photothermal,[9,10] and electrochemical means,[11,12] giving reduced GO (r-GO, a.k.a. chemically modified graphene). While r-GO is more defective and less conductive than pristine graphene produced by mechanical exfoliation, the comparatively easy synthesis and processing of GO make it an attractive precursor for large scale production of graphene-based materials.[13-15]

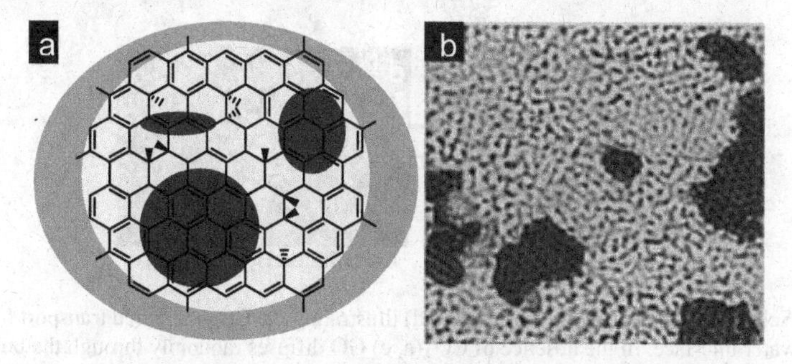

Figure 1. (a) Structural model of GO depicting ionizable hydrophilic edges (orange) and hydrophobic, unoxidized polyaromatic patches on the basal plane (blue). (b) Aberration-

corrected TEM image of the basal plane of a GO sheet, showing graphitic nanopatches (blue) surrounded by a matrix of oxidized material. (adapted from [20]).

GO is often described in the literature as hydrophilic due to its ability to disperse in water.[14, 16-19] However, while the edges of GO are hydrophilic, the basal plane still contains unoxidized π-conjugated islands nanometers in diameter, as highlighted in figure **1**.[20] Therefore, GO can be better described as an amphiphile, largely hydrophobic at the basal plane and hydrophilic at the edges. Armed with this hypothesis, we consider the various ways its amphiphilic behavior might be studied. Will GO resemble a small molecular amphiphile (dominated by its atom-scale thickness) or a colloidal surfactant (dominated by its micron-scale length and width)?[21] Can the hydrophobic-hydrophilic balance be tuned via pH, size, or degree of functionalization? What new surfactant functionalities might this unusual soft material have to offer?

Air-water surface activity

If GO is indeed a surfactant, we expect GO sheets to adhere to an air-water interface.[21] To detect GO sheets at the surface of water, we turn to Brewster angle microscopy (BAM), which images deviations in the refractive index contrast of the air-water surface.[22] As shown in figure **2a**, a BAM image of a freshly prepared GO dispersion detected little surface-active material. GO sheets began to populate the surface after several hours due to slow diffusion hindered by their large molecular mass. The transport of GO to the surface could be accelerated dramatically by introducing additional liquid-gas interfaces and/or allowing rising gas bubbles to carry GO sheets to the surface. Experimentally this was achieved by either bubbling air or nitrogen through the solution, or by preparing the GO dispersion from carbonated water and using boiling stones to liberate gas. This gave the BAM image of figure **2b**, where the water surface was coated with GO.[23]

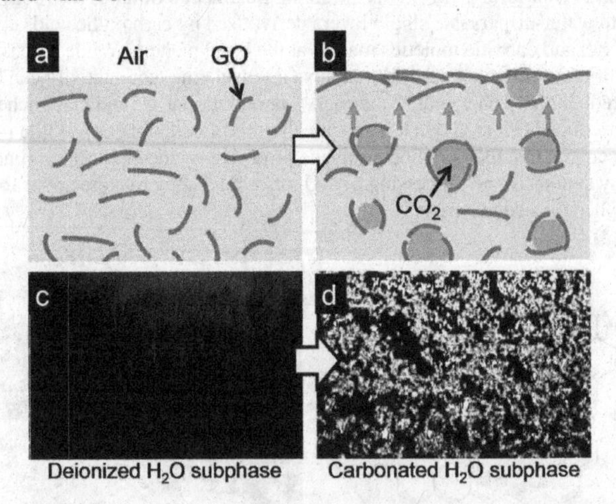

Figure 2. Schematic (a, b) and BAM images (c,d) illustrating the CO_2-mediated transport of GO to the air-water interface. In the absence of CO_2 (a, c) GO diffuses randomly through the bulk of the liquid, and takes several hours to populate the surface. When gas is bubbled through (b, d),

GO can be captured by rising bubbles and brought to the surface, giving dense surface coverage as indicated by a BAM image (d). (adapted from [23])

Oil-water interface

When a biphasic liquid of aqueous GO dispersion and toluene was shaken in a vial, GO acted as an emulsifier and created submillimeter droplets of the organic solvent dispersed in water which were stable for months. This behavior is characteristic of a Pickering emulsion,[24] suggesting GO is behaving as a colloidal surfactant.[21] As the concentration of GO was reduced, the size of the droplets increased while the volume of the emulsion decreased, as shown in figure **3**.[23] As with other particle stabilized emulsions, this emulsion is stable for several months owing to the large surface area of GO, which allows sheets to be kinetically trapped at the interface. Emulsion formation occurred more readily with aromatic solvents, like toluene and benzene, than with nonaromatic solvents, like chloroform and hexane, indicating that π-π interactions between the solvent and GO are a factor in emulsion stability.

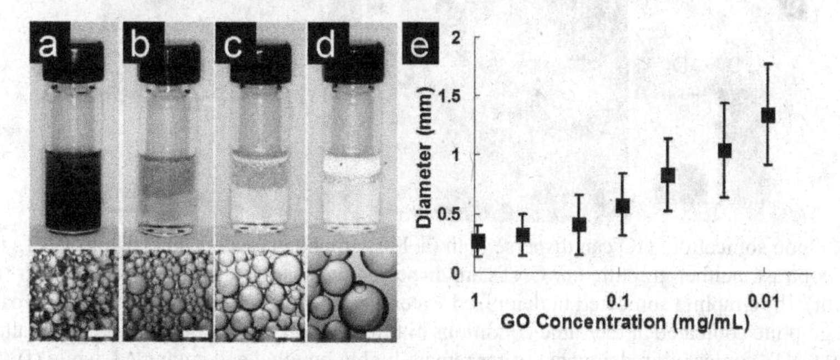

Figure 3. GO can form Pickering emulsions of aromatic oils in water. (a-d) Toluene droplets formed upon shaking a toluene/GO water mixture. The concentration of GO varies from (a) 0.95 mg/mL to (b) 0.19, (c) 0.047, and (d) 0.0095 mg/mL. The bottom row shows stereoscope images of the droplets in each case. All scale bars are 1 mm. (e) As GO concentration decreased, droplet diameter was observed to steadily increase. (adapted from [23])

Solid-water interface

Based on the GO's interfacial activities at air-water and oil-water interfaces, we proceeded to study its interfacial behaviors at solid-water interfaces, to see if GO could be used as a dispersing agent. To test the idea, we chose graphite and carbon nanotubes (CNTs) as model systems. These are notoriously difficult materials to process, and tremendous amounts of effort have been devoted to aqueous processing of CNTs using amphiphilic materials.[25-28] Indeed, graphite and CNTs each quickly settled after sonication in water, but both formed stable dispersions if sonicated in the presence of GO. As shown in figure **4a-c**, graphite powders sonicated in deionized water broke into platelets tens of microns across, while those sonicated in deionized water with GO broke into particles of only a few microns. This is likely because graphite pieces in the presence of GO are better suspended, and thus can be more effectively sonicated, than those without GO. Meanwhile, a heavily entangled multiwalled CNTs sample was dispersed and mostly disentangled by sonication with GO, while CNTs sonicated in deionized water remained aggregated and entangled (figure **4d-f**). Scanning electron microscopy

(SEM) and atomic force microscopy (AFM) images (figure **4f**) clearly showed that almost all CNTs in the sample were adhered to GO, indicating strong interactions between the two carbon materials[23]

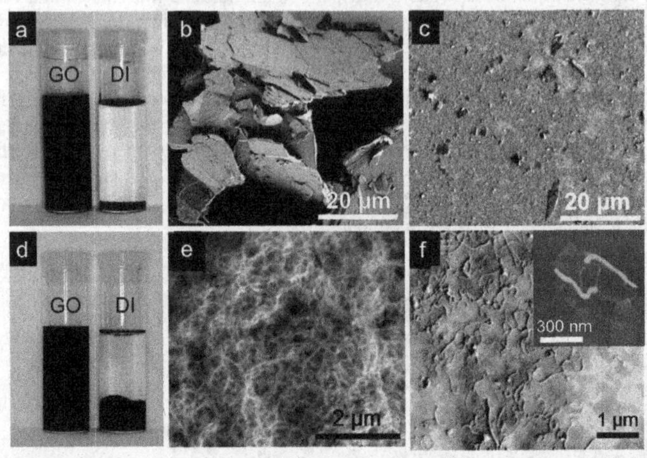

Figure 4. Upon sonication, GO can disperse both (a, left) graphite powders and (d, left) CNTs in water. In contrast, neither graphite nor CNTs are dispersed when sonicated in deionized water (a and d, right). (b) Graphite sonicated in deionized water breaks into pieces tens of microns across, while (c) graphite sonicated at the same conditions but in a GO dispersion breaks into far smaller pieces. (e) CNTs sonicated in deionized water remain highly bundled and entangled, while (f) those sonicated with GO are disentangled and well-dispersed. (f, inset) An AFM image shows that most CNTs are adhered to GO sheets after sonication. (adapted from [29])

Tunable amphiphilicity

The amphiphilic behavior of GO arises from the balance between hydrophilic groups at the edges and hydrophobic groups at the basal plane, and we can imagine several ways to tune that balance. Like that of ionic molecular surfactants, the amphiphilicity of GO can vary with the extent of ionization of the carboxylic acids at its edges, and thus can be modulated by changing the pH of the environment.[23,29] (figure **5a**) This is illustrated by changing the pH of the GO dispersion used to create the Pickering emulsion, as shown in figure **6a**. When the pH was lowered to 2 to fully protonate the carboxylates, GO was fully extracted from the water and into the emulsion because of its better balanced amphiphilicity at low pH. Meanwhile, when pH was raised to 10 to fully deprotonate the GO, making it very hydrophilic, the emulsion collapsed and GO returned to the aqueous phase. Surface tension measurements also confirmed that the surface activity of GO was most pronounced in the low pH state.[21]

Size too can affect GO's amphiphilicity. If the line density of carboxylates around the edge is roughly constant, then a smaller flake (with a larger edge-to-area ratio) will be more hydrophilic than a larger flake (with a smaller edge-to-area ratio).[29] This is represented schematically by figure **5b**. We demonstrate this by synthesizing "nano-GO" from commercially available graphite nanofibers rather than bulk graphite powders.[30] The nano-GO is more stable than its micron-scale counterpart at all pH values, according to zeta-potential measurements. Even nano-r-GO is somewhat stable against the aggregation to which regular-sized r-GO is prone. As shown in figure **6b**, CNTs dispersed in nano-GO formed a dispersion which not only was stable for months (as in the case of CNTs dispersed in regular GO), but which could withstand 10,000 RPM in a microcentrifuge without aggregating.

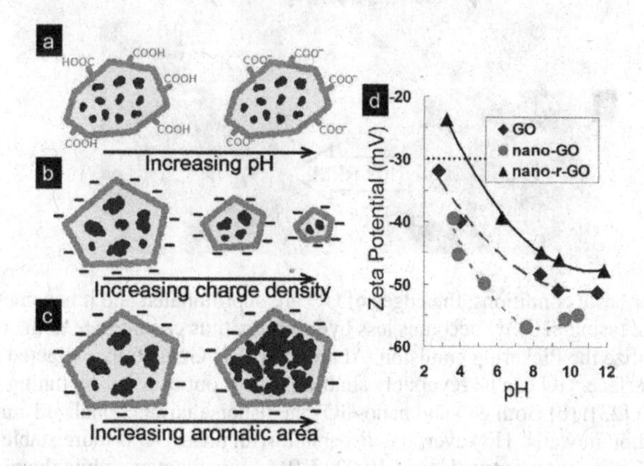

Figure 5. Amphiphilicity of GO can be tuned by pH, size, and degree of reduction, schematically shown in a, b, and c, respectively. (adapted from [29]) (d) Zeta-potential measurements clearly show that the charge density (hydrophilicty) of GO increases with higher pH and smaller size, but decreases when GO is reduced. (adapted from [30])

Lastly, the degree of functionalization at the hydrophobic basal plane can affect the amphiphilicity. With thermal or chemical reduction, the conjugated islands grow to occupy a greater portion of the basal plane area, and π-π interactions are strengthened. This is illustrated schematically in figure **5c** and reflected in the well-known tendency of dispersed GO to aggregate upon reduction.[6] Figure **5d** briefly plots the effects of all three variables upon the zeta-potential of GO.

Outlook

The ability of GO to revert to a conducting graphitic material upon reduction is well documented, and remains one of its most sought-after properties. GO's newly-discovered surfactant behavior offers another dimension of functionality, giving it extraordinary potential as an agent for aqueous processing of organic electronics. For instance, by constructing electric double-layer supercapacitors from the CNT-GO composite assembled by using GO as a surfactant, we generated a carbon scaffold with high specific surface area, giving a specific capacity of 175 F g^{-1} (as compared to 100 F g^{-1} in devices made from GO alone).[29] We have also used GO as a processing agent in an all-carbon bulk heterojunction active layer for a photovoltaic cell, which is described elsewhere.[31] In addition to the materials discussed here, we have observed that GO can disperse conjugated polymers such as polyaniline, and semiconducting small molecules such as perylene.

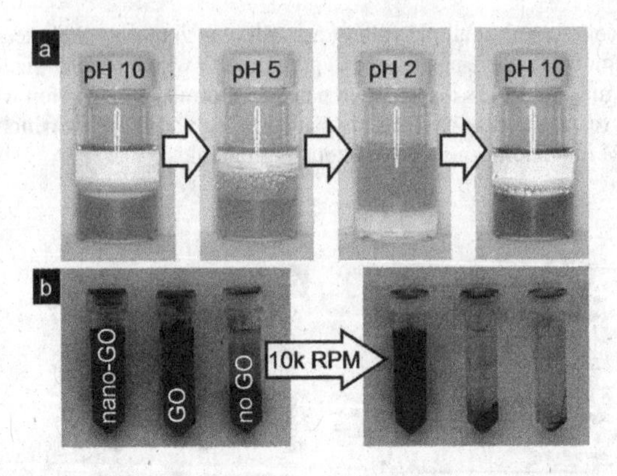

Figure 6. (a) Under basic conditions, the edges of GO are deprotonated and it becomes very hydrophilic. At decreasing pH, GO becomes less hydrophilic, thus can migrate to the oil-water interfaces and stabilize the Pickering emulsion. At around pH 2, GO is fully extracted from water to the oil-water interface. GO can be reversibly shuttled into or out of water by tuning pH in this way. (adapted from [23]) (b) Both GO and nano-GO can disperse unfunctionalized single-walled CNTs upon sonication in water. However, the dispersion with nano-GO is more stable in water.. The nano-GO-CNT dispersion can withstand 10,000 RPM centrifugation, while the regular GO dispersion cannot. (adapted from [29]).

In conclusion, we have demonstrated that GO, previously thought to be hydrophilic, is actually amphiphilic. GO is especially well suited to dispersion of aromatic hydrocarbons, and can disperse objects ranging from small molecules to micron-scale particles. GO's amphiphilicity and dispersion characteristics can be tuned by changing the pH of the environment or by varying the size or degree of oxidation of the GO itself. Because GO can be readily reduced to conductive r-GO, it can offer a bulk solution-processed route to organic composite materials with clean, electrically addressable interfaces.

Experimental details

GO was synthesized by a modified Hummer's method[1,2] from graphite powder purchased from Bay Carbon. Nano-GO was produced similarly from graphite nanofibers purchased from Catalytic Materials LLC. Multiwalled CNTs were purchased from Strem Chemicals and had an average diameter of 20 nm. For CO_2 flotation, GO was dispersed in commercially available carbonated water at a concentration of 0.01 mg/mL. Higher GO concentrations were found to hinder BAM by generating strong background scattering. The experiment was executed in a Langmuir-Blodgett trough equipped with a tensiometer and BAM (Nima Technology). Surface tension measurements are supported by drop shape analysis carried out on a Krüss DSA 100 instrument. Pickering emulsion experiments were carried out by mixing equal volumes of organic solvent and aqueous GO dispersion in a vial, and shaking by hand. Images were recorded directly through the vials using a Nikon SMZ-1500 stereoscope. Reported diameters represent the average measured diameters of >100 randomly chosen droplets from each experimental condition. pH was controlled by titrating with aqueous HCl or NaOH (each 1 M). Zeta-potential was measured with Malvern Instruments' Zetasizer Nano system. To produce aqueous dispersion of graphite particles or CNTs, they were added to GO dispersion at a mass ratio of 30:1 (graphite:GO) or 1:3 (CNT:GO), then sonicated for 30 min using a Misonix S-4000 cup-horn ultrasonicator. Samples were cooled with a chilled water pump to prevent thermal reduction of GO during sonication. The dispersion was centrifuged at 1,000 rpm for 5 min to remove undispersed material. The resulting dispersion was studied by SEM (Hitachi FE-SEM S-4800), and AFM (Veeco MultiMode V).

Acknowledgements

This work was supported by the National Science Foundation through a CAREER award (DMR 0955612). J.H. thanks the Northwestern Materials Research Science and Engineering Center (NSF DMR-0520513) for a capital equipment fund for the purchase of BAM and additional support from Sony Corporation. A.R.K. and L.J.C. are NSF graduate research fellows. J.H. is an Alfred P. Sloan Research Fellow. J.K.and J.L. gratefully acknowledges support from the Ryan Fellowship and the Northwestern University International Institute for Nanotechnology.

References

1. Cote, L. J., Kim, F., Huang, J., J. Am. Chem. Soc.,**131**, 1043 (2009).
2. Hummers, W. S. and Offeman, R. E., J. Am. Chem. Soc., **80**, 1339 (1958).
3. Lerf, A., He, H.Y., Forster, M., Klinowski, J., J. Phys. Chem. B **102**, 4477 (1998).
4. Gao, W., Alemany, L.B., Ci, L., Ajayan, P.M., Nat. Chem. **1**, 403 (2009).
5. Dobelle, W. H., Beer, M., J. Cell Biol. **39**, 733 (1968).
6. Stankovich, S., Dikin, D.A., Piner, R.D., Kohlhaas, K.A., Kleinhammes, A., Jia, Y., Wu, Y., Nguyen, S.T., Ruoff, R.S., Carbon **45**, 1558 (2007).
7. Croft, R.C., Quarterly Rev. **14**, 1 (1960).
8. Schniepp, H.C., Li, J.L., McAllister, M.J., Sai, H., Herrera-Alonso, M., Adamson, D.H., Prud'homme, R.K., Car, R., Saville, D.A., Aksay, I.A., J. Phys. Chem. B **110**, 8535 (2006).
9. Cote, L.J., Cruz-Silva, R., Huang, J.. J. Am. Chem. Soc. **131**, 11027 (2009).
10. Gilje, S, Dubin, S., Badakhshan, A., Farrar, J., Danczyk, S.A., Kaner, R.B., Adv. Mater. **22**, 419 (2010).
11. Wang, Z., Zhou, X., Zhang, J., Boey, F., Zha, H., J Phys. Chem. C **113**, 14071 (2009).
12. Ramesha, G.K., Sampath S., J. Phys. Chem. C **113**, 7985 (2009).
13. Allen, M.J., Tung, V.C., Kaner, R.B., Chem. Rev. **110**, 132 (2010).
14. Park, S., Ruoff R.S., Nat. Nanotechnol. **4**, 217 (2009).
15. Compton, O.C., Nguyen, S.T., Small **6**, 711 (2010).
16. Dikin, D.A., Stankovich, S., Zimney, E.J., Piner, R.D., Dommett, G.H.B., Evmenenko, G., Nguyen, S.T., Ruoff, R.S., Nature **448**, 457 (2007).
17. Gilje, S., Han, S., Wang, M., Wang, K.L., Kaner, R.B., Nano Lett. **7**, 3394 (2007).
18. Li, D., Kaner, R.B.. Science **320**, 1170 (2008).
19. Li, D., Muller, M.B., Gilje, S., Kaner, R.B., Wallace, G.G., Nat. Nanotechnol. **3**, 101 (2008).
20. Erickson K., Erni R., Lee Z., Alem N., Gannett W., Zettl A., Adv. Mater. **22**, 4467 (2010).
21. Myers D., Surfactant Science and Technology, Wiley-Interscience, Hoboken, NJ (2006).
22. Lipp, M.M., Lee, K.Y.C., Zasadzinski, J.A., Waring, A.J., Rev. Sci. Instrum. **68**, 2574 (1997).

23. Kim, J., Cote, L.J., Kim, F., Yuan, W., Shull, K.R., Huang, J., J. Am. Chem. Soc. **132**, 8180 (2010).
24. Pickering, S.U., J. Chem. Soc. **91**, 2001 (1907).
25. Islam, M.F., Rojas, E., Bergey, D.M., Johnson, A.T., Yodh, A.G., Nano Lett. **3**, 269 (2003).
26. Moore, V.C., Strano, M.S., Haroz, E.H., Hauge, R.H., Smalley, R.E., Schmidt, J., Talmon, Y., Nano Lett. **3**, 1379 (2003).
27. Grossiord, N., Loos, J., Regev, O., Koning, C.E., Chem. Mater. **18**, 1089 (2006).
28. Vaisman, L., Wagner, H.D., Marom, G., Adv. Colloid Interface Sci. **128**, 37 (2006).
29. Cote, L.J., Kim, J., Tung, V.C., Luo, J., Kim, F., Huang, J., Pure Appl. Chem. **83**, 1, 95 (2011).
30. Luo, J., Cote, L.J., Tung, V.C., Tan, A.T.L., Goins, P.E., Wu, J., Huang, J., J. Am. Chem. Soc., **132**, 17667 (2010).
31. Tung, V.C., Huang, J.H., Tevis, I., Kim, F., Kim, J., Chu, C.W., Stupp, S.I., Huang, J., J. Amer. Chem. Soc., **133**, 4940 (2011).

Mater. Res. Soc. Symp. Proc. Vol. 1344 © 2011 Materials Research Society
DOI: 10.1557/opl.2011.1358

The Dynamics of Formation of Graphane-like Fluorinated Graphene Membranes (Fluorographene): A Reactive Molecular Dynamics Study

Ricardo P. B. Santos[1,2], Pedro A. S. Autreto[1], Sergio B. Legoas[3], and Douglas S. Galvao[1]
[1] Instituto de Física "Gleb Wataghin, Universidade Estadual de Campinas, Campinas - SP, 13083-970, Brazil
[2] Universidade Estadual de Maringá, 82020-900, Maringá - PR, Brazil.
[3] Departamento de Física, CCT, Universidade Federal de Roraima, 69304-000, Boa Vista - RR, Brazil.

ABSTRACT

Using fully reactive molecular dynamics methodologies we investigated the structural and dynamical aspects of the fluorination mechanism leading to fluorographene formation from graphene membranes. Fluorination tends to produce significant defective areas on the membranes with variation on the typical carbon-carbon distances, sometimes with the presence of large holes due to carbon losses. The results obtained in our simulations are in good agreement with the broad distribution of values for the lattice parameter experimentally observed. We have also investigated mixed atmospheres composed by H and F atoms. When H is present in small quantities an expressive reduction on the rate of incorporation of fluorine was observed. On the other hand when fluorine atoms are present in small quantities in a hydrogen atmosphere, they induce an increasing on the hydrogen incorporation and the formation of locally self-organized structure of adsorbed H and F atoms.

INTRODUCTION

Graphene is a two dimensional array of hexagonal units of sp^2 bonded C atoms [1]. Because of its electronic properties, it is considered one of the most promising materials for future electronics [2]. However, in its pristine state, it is a gapless material, which poses serious limitations to a series of electronic applications [3]. The production of graphene-like structures with well defined gap has been tried using different approaches. One possibility towards opening graphene gap is through chemical functionalization, using hydrogen or fluorine atoms [4-13].

Fully hydrogenated graphene, named graphane, was theoretically predicted by Sofo, Chaudhari, and Barber [4], and experimentally realized by Elias *et al.* [6]. In their experiments, graphene membranes were submitted to H^+ exposure from cold plasma. The H incorporation into the membranes results in altering the carbon hybridizations from sp^2 to sp^3.

A topological similar structure can be produced using fluorine atoms. Perfect idealized fluorographene would consist of a single-layer structure with fully saturated (sp^3 hybridization) carbon atoms and with C-F bonds in an alternating pattern (up and down with relation to the plane defined by the carbon atoms) , these patterns are predicted to be found in chair, zig-zag, boat or armchair-like structures [10].

MODELING

We carried out molecular dynamics (MD) simulations in order to investigate the structural and dynamical aspects of the hydrogenation and fluorination of graphene membranes, leading to the formation of graphane and flurographene-like structures.

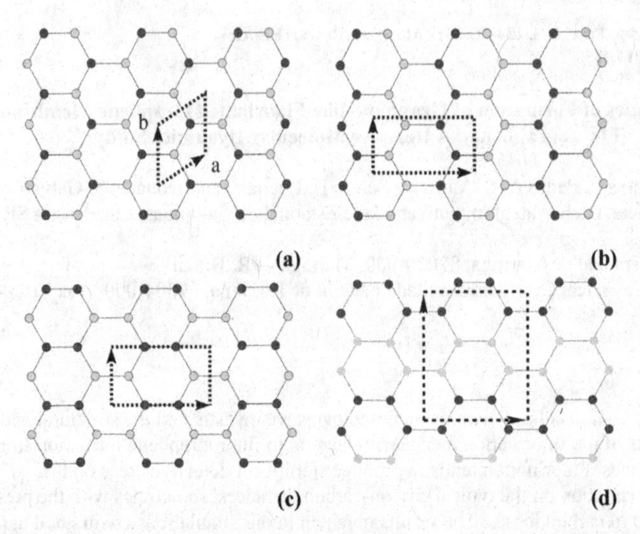

Figure 1: Structural models of four possible configurations of fluorographene. Fluorine atoms are indicated in dark gray (blue online) if adsorbed over the carbon membrane and in light gray (yellow online) if adsorbed below it. The structures are named: (a) chair-like , (b) zig-zag-like, (c) boat-like and (d) armchair-like, respectively.

The extensive MD simulations were carried out using reactive force fields (ReaxFF [14-16]), as implemented in the Large-scale atomic/Molecular Massively Parallel Simulator (LAMMPS) code [17]. Graphene membranes, immersed into different atmosphere (composed of pure and mixed H and F atoms) and temperature conditions were considered. The temperature range was from T=300 K up to T=650 K, while the atmospheres considered ranged from ratios of F/C and H/C atoms varying from 0.1 to 2.0. We have also considered also the cases of mixed F and H atmospheres, in order to verify how the F and H incorporation rates would be affected. Finite structures with H-passivated borders were considered in the simulations, and the membranes were embedded into atmospheres of randomly distributed atoms (H and/or F). Typical membrane structures contain about 10,000 carbon atoms. In order to speed up the simulations, the H-H and F-F recombinations were not allowed during the simulations. A Langevin thermostat, as implemented in LAMMPS code, was used and the typical time for a complete simulation run was 500 ps, with time-steps of 1fs.

ReaxFF is a reactive force field developed by van Duin, Goddard III and co-workers for use in MD simulations. It allows simulations of many types of chemical reactions. It is similar to standard non-reactive force fields, like MM3 [14-16], where the system energy is divided into partial energy contributions associated with, amongst others, valence angle bending and bond stretching, and non-bonded van der Waals and Coulomb interactions [14-16]. However, one main difference is that ReaxFF can handle bond formation and dissociation (making/breaking bonds) as a function of bond order values. ReaxFF was parameterized against DFT calculations, being the average deviations between the heats of formation predicted by the ReaxFF and the experiments equal to 2.8 and 2.9 kcal/mol, for non-conjugated and conjugated systems, respectively [14-16]. We have carried out geometry optimizations using gradient conjugated techniques (convergence condition with gradient values less than 10^{-3}).

RESULTS AND DISCUSSIONS

For the simulations with F/C rates near to 1.0, it was possible to observe that at high temperatures (500 K and above), despite of obtaining higher incorporation rates for all simulations, fluorine strong reactivity caused extensive damages to the membrane for long simulation times. For these reasons, we will only discuss in details the results for 300 K.

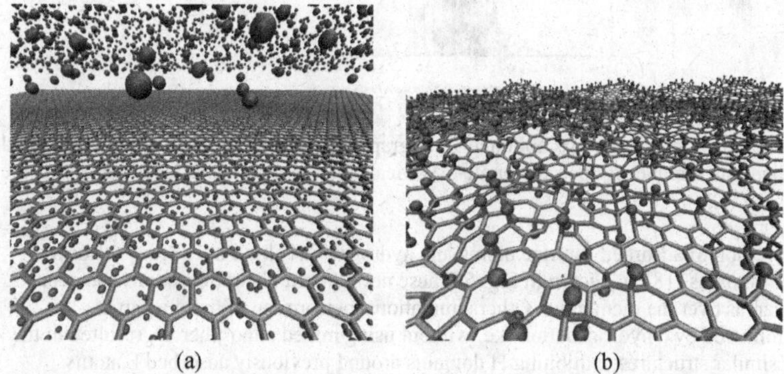

Figure 2: Representative snapshots from MD simulations of the fluorination process for initial (a) and final stage (b) configurations.

In Fig. 2 we present representative snapshots from the early and final fluorination stages of a simulation carried out at 300K. At the beginning we have a graphene membrane in a planar configuration immersed into a fluorine atmosphere (Fig. 2a). F incorporation generates a corrugated structure as showed in the Fig. 2b. The results showed that the initial incorporations are made preferably of chair-like (*trans*-configuration (Fig. 2b) structures. Also, a significant large number of uncorrelated small F domains are formed in the early stages. These results also suggest that large domains of perfect fluorographene-like structures are unlikely to be formed.

The fluorine incorporation creates large corrugated areas on the membrane, and affects the typical C-C distances and the lattice parameter as well. These deformations are directly related with the change in the distribution of second-neighbor distances (lattice parameter) on the membrane, and can be seen in Fig. 3. The broadening of the distribution after fluorination and the increase in the mean value of second-neighbor distances (~1.3%) are in good agreement with the results reported by Geim and co-workers [12].

We have also carried out simulations considering a fluorine atmosphere containing a small quantity of hydrogen atoms. We observed that in this case the net effect was to decrease the fluorination rate. For the opposite case, a hydrogen atmosphere containing a small quantity of fluorine atoms, it is possible to observe a remarkable increase on the rate of hydrogen incorporation (catalytic-like effect) after some time. This time was of the order of 200 ps, as shown in Fig. 4, and represents the necessary amount of time to the number of incorporated fluorine atoms be enough to create many reactive sites on the membrane.

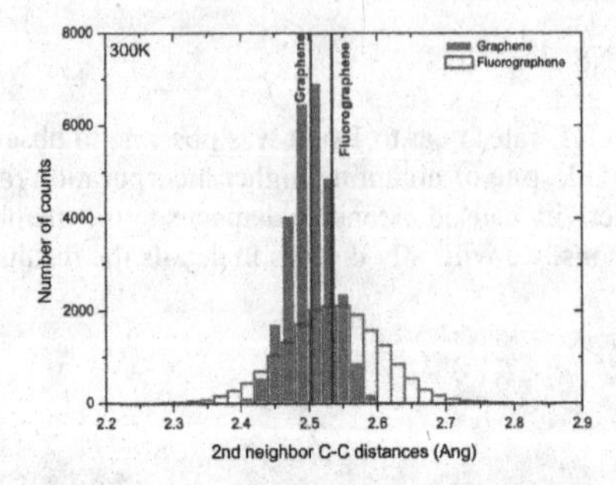

Figure 3: Comparison of histograms obtained for a graphene membrane in filled rectangles and a fluorographene membrane in open rectangles. Vertical lines represent the mean values of the distributions.

In the region of adsorbed fluorine atoms, the hydrogen ones have a tendency to form graphane-like domains [18], as shown in Fig. 5. These new structural configurations can occur for extensive areas over the membrane. Other simulations, performing initially a partial fluorination followed by a hydrogenation (i.e. without using mixed atmospheres), resulted in the formation of similar structures, exhibiting H domains around previously adsorbed F atoms.

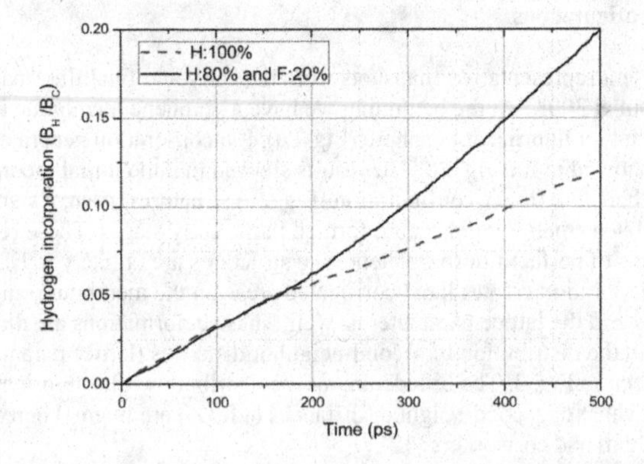

Figure 4: Hydrogen incorporation in the case of pure (H) and mixed (80% of H atoms and 20% of F atoms) atmospheres. The catalytic-like effect of introducing fluorine atoms is clear after 200ps of simulation.

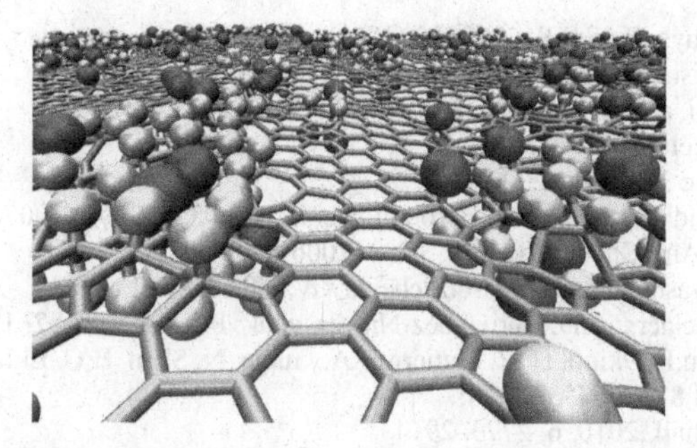

Figure 5: H islands of adsorbed H atoms formed in the vicinity of adsorbed F atoms, creating mixed F and H structures. F atoms are represented in dark gray (red online) and H atoms are light gray (yellow online), while the hexagonal network represents the carbon membrane.

SUMMARY AND CONCLUSIONS

We have investigated using reactive molecular dynamics simulations the fluorination mechanisms leading to fluorographene formation from graphene membranes. Our results showed changes on the typical carbon-carbon distances and cell parameter values consistent with values experimentally reported. Our simulations also showed that the chair-like configuration is the most likely to occur in the formation of fluorographene. The simulations with mixed atmospheres showed two different behaviors, one with F atmospheres with small quantities of hydrogen atoms (decreasing the F incorporation rate) and other with H atmospheres containing small quantities of fluorine atoms (producing a catalyst-like effect on the hydrogenation rate). Also, we observed the formation of mixed structures, with H and F atoms adsorbed in self-organized domains for both mixed atmospheres (H and F), as well as, for fluorination followed by hydrogenation. It is expected that these stable configurations will present a well-defined non-zero bandgap values with potential use in nanoelectronics.

ACKNOWLEDGMENTS

This work was supported in part by the Brazilian agencies CNPq, CAPES and FAPESP. The authors wish to thank Professor A. van Duin for his very helpful assistance with ReaxFF code.

REFERENCES

[1] K. S. Novoselov *et al.*, Science **306**, 666 (2004).

[2] S. H. Cheng *et al.*, Phys. Rev. B **81**, 205435 (2010).
[3] F. Withers, M. Duboist, and A.K. Savchenko, arxiv:1005.3474v1 (2010).
[4] J. Sofo, A. Chaudhari, and G. Barber, Phys. Rev. B **75**, 153401 (2007).
[5] S. Ryu *et al.*, Nano Lett. **8**, 4597 (2008).
[6] D. Elias *et al.* Science **323**, 610 (2009).
[7] J. O. Sofo, A. S.Chaudhari, and, G. D. Barber, Phys. Rev. B **75**, 153401 (2007).
[8] D. Lueking *et al.*, J. Am. Chem. Soç. **128**, 7758 (2006).
[9] N. R. Ray, A. K. Srivastava, and, R. Grotzsche, arXiv:0802.3998v1 (2008).
[10] O. Leenaerts, H. Peelaers, A.D. Hernandez-Nieves, et al.; Phys. Rev. B 82, 195436 (2010).
[11] S.-H. Cheng, K. Zou,F. Okino, H. R. Gutierrez, A. Gupta, N. Shen, P. C. Eklund, J. O. Sofo, and J. Zhu, Phys. Rev. B **81**, 205435 (2010).
[12] R. R. Nair *et al.*, Small, 2010, 6, 2773–2914.
[13] J. T. Robinson *et al.*, Nano Lett., *in press*, DOI: 10.1021/nl101437p.
[14] A. C. T. van Duin, S. Dasgupta, F. Lorant, and W. A. Goddard III, J. Phys. Chem. A **105**, 9396 (2001).
[15] A. C. T. van Duin and J. S. S. Damste, Org. Geochem. **34**, 515 (2003).
[16] K. Chenoweth, A. C. T. van Duin, and W. A. Goddard III, J. Phys. Chem. A **112**, 1040 (2008).
[17] S. Plimpton, J. of Comp. Phys., 117, 1–19 (1995).
[18] M. Z. S. Flores, P. A. S. Autreto, S. B. Legoas, and D. S. Galvao, Nanotechnology **20**, 465704 (2009).

Mater. Res. Soc. Symp. Proc. Vol. 1344 © 2011 Materials Research Society
DOI: 10.1557/opl.2011.1353

DNA Gating effect from single layer graphene

Jian Lin[1,2], Desalegne Teweldebrhan[2], Khalid Ashraf[2], Guanxiong Liu[2], Xiaoye Jing,[2] Zhong Yan[2], Mihrimah Ozkan[2], Roger K. Lake[2], Alexander A. Balandin[2,3], Cengiz S. Ozkan[1,3]

[1]Department of Mechanical engineering, [2] Department of Electrical engineering, [3] Department of Material science and engineering, University of California at Riverside, Riverside, CA 92521, U.S.A.

ABSTRACT

In this letter, single stranded Deoxyribonucleic Acids (ssDNA) are found to act as negative potential gating agents that increase the hole density in single layer graphene (SLG). Current-voltage measurement of the hybrid ssDNA/graphene system indicates a shift in the Dirac point and "intrinsic" conductance after ssDNA is patterned. The effect of ssDNA is to increase the hole density in the graphene layer, which is calculated to be on the order of 1.8×10^{12} cm^{-2}. This increased density is consistent with the Raman frequency shifts in the G-peak and 2D band positions and the corresponding changes in the G-peak full-width half maximum. This patterning of DNA on graphene layers could provide new avenues to modulate their electrical properties and for novel electronic devices.

INTRODUCTION

Since isolated in 2004[1], graphene emerged as a potential material for fabricating nanoelectronics beyond CMOS. Deoxyribonucleic acid (DNA) and peptide nucleic acids (PNAs) which have base sequences that offer specificity are attractive assembly linkers for bottom-up nanofabrication. Engineered ssDNA sequences are employed in the nanoarchitectures of end-functionalized single walled carbon nanotubes for device applications including resonant tunneling diodes, field effect transistors and biochemical sensors [2-4]. A thorough understanding of electrical transport through the interface between biological molecules such as DNA and graphene layers is still in its infancy[5]. In this work, we investigate the modulation of carrier transport through graphene layers with overlaying ssDNA fragments via electrostatic interaction. We discovered that the role of ssDNA on the surface of graphene is analogous to applying a negative gate potential in conventional silicon CMOS architectures. Raman spectroscopy has been used in monitoring the doping of graphene layers [6-8].

EXPERIMENT

The graphene samples used in this study were extracted from highly oriented pyrolitic graphite (HOPG) slabs via mechanical exfoliation. The graphene samples were deposited on p-type degenerately doped Si (100) wafers (p++) covered with 300nm thick thermally grown SiO_2. For device fabrication, we patterned source and drain contacts using electron beam lithography followed by the deposition of source and drain metal consisting of 10nm thick Cr and 100nm thick Au layers (Temescal BJD-1800 electron beam evaporator).

The ssDNA sequence employed in this work is CGGGAGCTCAGCGGATAGGTGGGC. The engineered oligonucleotides (Sigma Genosys) were diluted in distilled water to obtain the stock solution. Concentration of the ssDNA solution was calculated to be 28.86mg/ml. After the current-voltage measurements (I_{ds}-V_{gs}) of as-fabricated SLG transistors were measured in vacuum (1.0×10^{-3} Torr) at 300K, a 0.5μl droplet of ssDNA solution was patterned over the active part of the device, followed by incubation for 30 minutes in ambient environment and nitrogen drying. Then the current-voltage measurement of SLG transistors with patterned ssDNA was operated under vacuum (1.0×10^{-3} Torr) at 300K.

The Raman microscopy was used to identify the single layer graphene as well as the the Raman peak position shift before and after patterning the ssDNA on the top of the SLG. The Raman microscopy was carried out using the Renishaw instrument. The spectra were excited by the 488 nm visible laser. A Leica optical microscope with 50x objective was used to collect the backscattered light from the graphene samples. The Rayleigh light was rejected by the holographic notch filter with a 160 cm^{-1} cut off frequency for 488 nm excitation. The spectra were recorded with the 1800 lines/mm grating. A special precaution was taken to avoid the local heating of the samples by the excitation laser. In order to achieve this, all measurements were carried out at low excitation power, below 2 mW on the sample surface. The power density on the sample surface was verified with an Orion power meter. The spectral resolution of the instrument determined by the hardware was ~1 cm^{-1}. The spectral resolution enhanced by the software processing of the peak positions was below 0.5 cm^{-1}.

DISCUSSION

A statistical study based on atomic force microscopy (AFM) imaging of the SLG layer before and after ssDNA patterning is shown in Figure 1. The average thickness of the SLG layer increased from ~1.0nm to ~1.8nm, indicating the formation of a thin layer of ssDNA on top of the SLG surface. From the AFM images, we observe the increased roughness of both the SiO_2 substrate and the SLG surface, which could be due to nonspecific and irregular binding of ssDNA fragments.

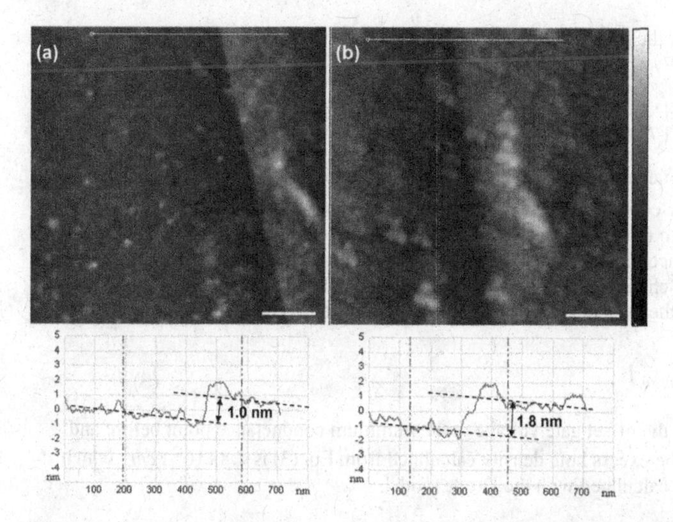

Figure 1. (a), (b) AFM images (1.1 um×1.1um, Z-range 15nm) and accordingly cross sectional profiles of SLG before (a) and after (b) patterning with ssDNA. The measured thickness of SLG increases from ~1.0nm to ~1.8nm after ssDNA patterning. And the increased roughness (shown in two squares in image (a) and (b)) is due to the nonspecific and irregular binding of ssDNA to the SLG. The scale bar is 200nm. Color bar is 10nm.

Current-voltage measurements on the SLG transistor are conducted in vacuum (1.0×10^{-3} Torr) before and after ssDNA deposition to determine the modulation of electrical characteristics. Employing a simple MOSFET model [9], the carrier mobility was calculated using the following equation,

$$\mu = \left(\frac{\Delta I_{ds}}{\Delta V_{gs}}\right) / \left(\frac{C_g W V_{ds}}{L}\right) \tag{1}$$

where I_{ds} and V_{ds} are the source-drain current and voltage, V_{gs} is the back-gate source voltage, L and w are the effective channel length and width respectively, and C_g is the gate capacitance of the SLG device. Based on the data shown in Figure 2, the carrier mobility of the SLG transistor at the room temperature before ssDNA patterning was calculated to be ~1303 $cm^2/(V\cdot s)$. From Figure 2, we observe the minimum conductance point (when the Fermi level is at the Dirac point) and the corresponding finite offset gate voltage which depends on the charge impurity[10]. After ssDNA is patterned on the SLG devices, a parallel shift in the minimum conductance point (MCP) is noticed, which may originate from the creation of more charge impurities. In other words, when negatively charged ssDNA attaches to graphene, we need to apply more positive voltage to compensate for this additional charge[11]. In a control experiment, we placed a droplet of DI water on the SLG devices. Then after drying via nitrogen flow, electrical measurement under the vacuum was conducted, and no such shift in the MCP was observed. The increase in the offset of the Dirac point gate voltage is around 25V. The conductance of the SLG transistor is also increased; the increased hole density induced via ssDNA patterning was calculated to be $3.45\times10^{12}\,cm^{-2}$ using the Drude model,

$$\Delta p = \left(R_2^{-1} - R_1^{-1}\right) / \left(\mu_p q (\frac{W}{L})\right) \tag{2}$$

Where R_2, R_1 are the resistances of the SLG device before and after ssDNA patterning. This increase in the hole density could be attributed to the following reason: ssDNA molecules with negative charge could act to induce a negative electric field effect (EFE) on the SLG[5]. This EFE could effectively induce the injection of extra holes in the graphene layer, hence the conductivity changes. The change in the hole density is also calculated from a parallel plate capacitor model based on the shift in the gate voltage of the minimum conductance point (MCP),

$$\Delta p = \frac{C_g}{q}(V_{mcp2} - V_{mcp1}) \tag{3}$$

where V_{mcp2} and V_{mcp1} are the offset gate voltage in the minimum conductance point before and after ssDNA patterning. The excess hole density calculated from Eq. (3) is $1.8\times10^{12}\,/\,cm^2$ which is comparable to the value calculated with the Drude model.

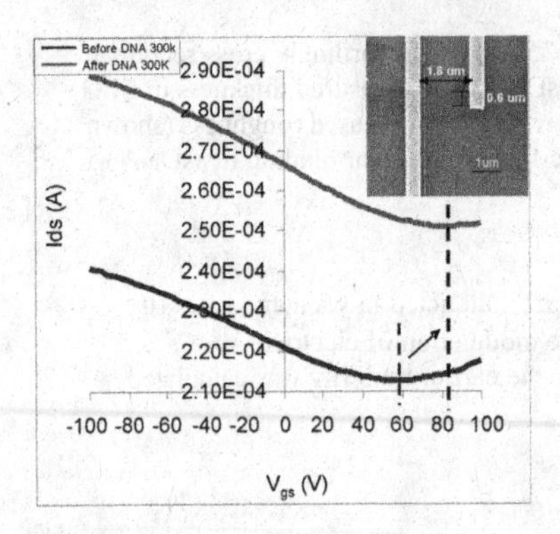

Figure 2. I_{ds}-V_{gs} characteristics of SLG before (black) and after (red) ssDNA patterning of SLG devices at T=300K. Source-drain bias voltage is 100mV. The doted lines correspond to the Dirac point (Minimum conductance point). From the current-voltage measurements, it shows: 1) the gate voltage when minimum conductance point reaches shifts from 60V to 80V; 2) overall conductance shifts up. The top inset is the SEM image of the SLG device used in the measurements. The device channel length and effective width are 1.8um and 0.6um respectively.

Raman spectroscopy has been intensively used for identifying the number of layers[12] and monitoring the doping of graphene layers. The major Raman spectra features for graphene are the G band (optical phonon at long wavelengths) at ~1584cm^{-1}, and the 2D band (associated with a two-phonon state) at ~2700cm^{-1} in pristine SLG. However, in doped graphene, the G band and 2D band frequencies as well as the full width at half maximum (FWHM) value in the G peak are changed [6-8]. To validate that ssDNA acting as a negative potential gating agent on top of SLG can induce additional holes, we measured the Raman spectrum on pristine SLG samples without device fabrication which may induce some contaminants and residue on the surface [13]. Next we patterned the same concentration of ssDNA on top of the SLG samples. Then the same Raman measurement was operated on the SLG samples with patterned ssDNA. The constant laser power below 2mW was employed to avoid the flakes and the patterned ssDNA layer are not damaged and to rule out the local heating effects. In Figure 3, we present the Raman spectra for which the main feature is the shift in the G and the 2D band positions as well as their FWHM. After ssDNA patterning, the average position of the G peak shifts from 1580cm^{-1} to 1582cm^{-1}. According to measurements by Ferrari et al.,[6] a shift of ~2cm^{-1} corresponds to a hole concentration of approximately $2.0\times10^{12} / cm^2$, which is comparable to the value based on electrical measurements. The reduction of the FWHM of the G peak after negative ssDNA gating is similar to the previously reported results for conventional electrostatic gating [6-8,14,15]. A detailed explanation for the decrease in the phonon G-peak FWHM value with an increase in carrier concentration has been previously provided [6-8].

Apart from the interpretation of the G peak, the 2D peak originates from a second-order, double-resonant (DR) Raman scattering mechanism[14]. We measured the shift of the 2D peak position after ssDNA patterning. In our measurements, the 2D band line is averagely centered at 2685cm^{-1} before ssDNA patterning and shifted to 2689cm^{-1} after ssDNA patterning, which is in agreement with the previously obtained results [7,15]. The observed decrease in the FWHM in the 2D band could be due to the fact that the electrostatic interaction between DNA and SLG could further modify the 2D band behavior.

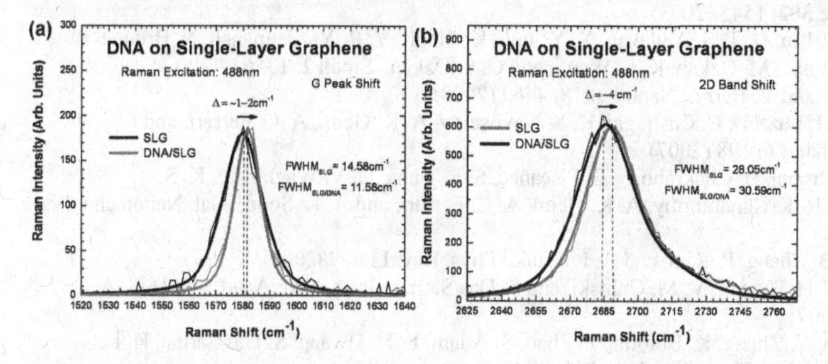

Figure 3. Micro-Raman spectroscopy of the signature G-peak and the 2D band for pristine SLG and the ssDNA/SLG system. In (a) the G-peak is centered at 1580cm^{-1}, observed at 488nm excitation wavelength, shifts an average 2cm^{-1} after ssDNA patterning. (b) Second order 2D band shifts by ~4cm^{-1} as a result of ssDNA patterning. Both signature peaks also show relative shortening of their respective full width half max (FWHM) values.

CONCLUSIONS

We demonstrated that ssDNA fragments act as negative potential gating agents resulting in an increase in the hole density in graphene layers. By using current-voltage measurements, we computed an increase in the hole density with a value of about 1.8×10^{12} cm^{-2} based on both the change in resistance and the shift in the voltage of the minimum conductance point. This value is consistent with the peak-position shift of the G band and the 2D band of the Raman spectra. Furthermore, we discussed the relationships between the gating effect induced by the ssDNA fragments and the change in the phonon frequency, and demonstrated that patterning of biomolecules on graphene layers could provide new avenues to modulate their electrical properties.

ACKNOWLEDGMENTS

We gratefully acknowledge funding for this work by the FCRP Center on Functional Engineered Nano Architectonics (FENA), the Nanomanufacturing Division of the National Science Foundation (Award Number: 0800680) and the Center for Hierarchical Manufacturing (CHM) funded by the NSF.

REFERENCES

1. K. S. Novoselov, A. K. Geim, S. V. Morozov, D. Jiang, Y. Zhang, S. V. Dubonos, I. V. Grigorieva, and A. A. Firsov, Science **306**, 666 (2004).
2. R. R. Johnson, A. T. C. Johnson, and M. L. Klein, Nano Lett. **8**, 69 (2008).
3. M. Zheng, A. Jagota, M. S. Strano, A. P. Santos, P. Barone, S. G. Chou, B. A. Diner, M. S. Dresselhaus, R. S. McLean, G. B. Onoa, G. G. Samsonidze, E. D. Semke, M. Usrey, and D. J. Walls, Science **302**, 1545 (2003).
4. X. Wang, F. Liu, G. T. S. Andavan, X. Y. Jing, K. Singh, V. R. Yazdanpanah, N. Bruque, R. R. Pandey, R. Lake, M. Ozkan, K. L. Wang, and C. S. Ozkan, Small **2**, 1356 (2006).
5. N. Mohanty and V. Berry, Nano Lett. **8**, 4469 (2008).
6. S. Pisana, M. Lazzeri, C. Casiraghi, K. S. Novoselov, A. K. Geim, A. C. Ferrari, and F. Mauri, Nat. Mater. **6**, 198 (2007).
7. A. Das, S. Pisana, B. Chakraborty, S. Piscanec, S. K. Saha, U. V. Waghmare, K. S. Novoselov, H. R. Krishnamurthy, A. K. Geim, A. C. Ferrari, and A. K. Sood, Nat. Nanotechnol. **3**, 210 (2008).
8. J. Yan, Y. B. Zhang, P. Kim, and A. Pinczuk, Phys. Rev. Lett. **98**2007).
9. S. Adam, E. H. Hwang, V. M. Galitski, and S. Das Sarma, Proc. Natl. Acad. Sci. U. S. A. **104**, 18392 (2007).
10. Y. W. Tan, Y. Zhang, K. Bolotin, Y. Zhao, S. Adam, E. H. Hwang, S. Das Sarma, H. L. Stormer, and P. Kim, Phys. Rev. Lett. **99**2007).
11. A. B. Artyukhin, M. Stadermann, R. W. Friddle, P. Stroeve, O. Bakajin, and A. Noy, Nano Lett. **6**, 2080 (2006).
12. A. C. Ferrari, J. C. Meyer, V. Scardaci, C. Casiraghi, M. Lazzeri, F. Mauri, S. Piscanec, D. Jiang, K. S. Novoselov, S. Roth, and A. K. Geim, Phys. Rev. Lett. **97**2006).
13. M. Ishigami, J. H. Chen, W. G. Cullen, M. S. Fuhrer, and E. D. Williams, Nano Lett. **7**, 1643 (2007).
14. C. Casiraghi, S. Pisana, K. S. Novoselov, A. K. Geim, and A. C. Ferrari, Appl. Phys. Lett. **91**, 233108 (2007).
15. C. Stampfer, F. Molitor, D. Graf, K. Ensslin, A. Jungen, C. Hierold, and L. Wirtz, Appl. Phys. Lett. **91**, 241907 (2007).

Mater. Res. Soc. Symp. Proc. Vol. 1344 © 2011 Materials Research Society
DOI: 10.1557/opl.2011.1356

Data Transmission Performance of Few-Layer Graphene Ribbons

Ali Bilge Guvenc[1], Jian Lin[2], Miroslav Penchev[1], Cengiz Ozkan[2, 3], Mihrimah Ozkan[1]
[1]Department of Electrical Engineering, University of California-Riverside, Riverside, CA 92521, U.S.A.
[2]Department of Mechanical Engineering, University of California-Riverside, Riverside, CA 92521, U.S.A.
[3]Material Science and Engineering Program, University of California-Riverside, Riverside, CA 92521, U.S.A.

ABSTRACT

We investigated the electrical characteristics and digital data transmission performance few-layer graphene ribbons grown by chemical vapor deposition. Graphene ribbons having a mobility of 2,180 $cm^2V^{-1}s^{-1}$ can sustain data rates up to 50 megabits per second at 1.5 μm length, thus the bandwidth is inversely proportional to resistance caused by defects in the graphene layers. Improving the graphene mobility to highest measured values (~200,000 $cm^2V^{-1}s^{-1}$) and using structures with multiple coplanar transmission lines in parallel could carry the bandwidth beyond the gigabits per second level.

INTRODUCTION

As the length scale of devices become smaller,[1] the dimensions of integrated circuits (IC) steadily decrease. Using metal interconnects between the devices leads to performance limitations because their electrical properties and mechanical stability reach their theoretical limits[2] beyond the 130 nm technology node. One promising candidate which has been proposed to out-perform copper (Cu) is the graphene ribbons[3] (GRs) which can be applied with the current technology without changing the top-down fabrication methods. Graphene is a single atomic layer of sp^2-bonded carbon atoms packed into a honeycomb lattice. Since its first discovery in 2004,[4] it has drawn great attention due to its remarkable properties such as linear energy dispersion relation[5] and ultrahigh charge carrier mobility at room temperature (~200,000 $cm^2 V^{-1}s^{-1}$).[6]

Previous studies on the performance of graphene ribbons have mostly focused on the electrical and thermal properties of the material. In this work, we report on the basic electrical transport characterization, high-speed digital data transmission performance measurements and the analysis of GRs by using the eye diagram[7] of the transmission line. This method allows the determination of the performance parameters of the line including the bit error rate (BER), the quality (Q) factor[8], signal attenuation and maximum bandwidth[9].

EXPERIMENT

The devices used in this work were fabricated from large area few-layer graphene sheets grown by chemical vapor deposition and patterning the sheets in the form of stripes via lithographic patterning (Figure 1a and 1b). Large area graphene sheets were grown on thin Ni catalyst films using highly diluted methane ($Ar:H_2:CH_4$=600:500:30 standard cubic centimeter per minute (sccm)) at 900 °C under ambient pressure conditions. The 300 nm thick Ni films were deposited on Si/SiO_2 substrates using electron beam evaporation, followed by heating at 1,000 °C under Ar/H_2 (600:500 sccm) atmosphere in ambient pressure for 30 minutes to form a polycrystalline Ni film with large grain size.[10] The furnace was then cooled to 900 °C at rate of

50 °C/min. After flowing 30 sccm methane mixed with Ar/H_2 at 900 °C under ambient pressure for 1 minute, the temperature was cooled down to 25°C at medium/fast cooling rate of 400 °C/min. After the growth was completed, a mild HCl aqueous solution (3%) was used to etch the residual Ni film underneath the graphene layers. After that, the graphene layers were transferred onto a Si/SiO_2 substrate for device fabrication and characterization.[11]

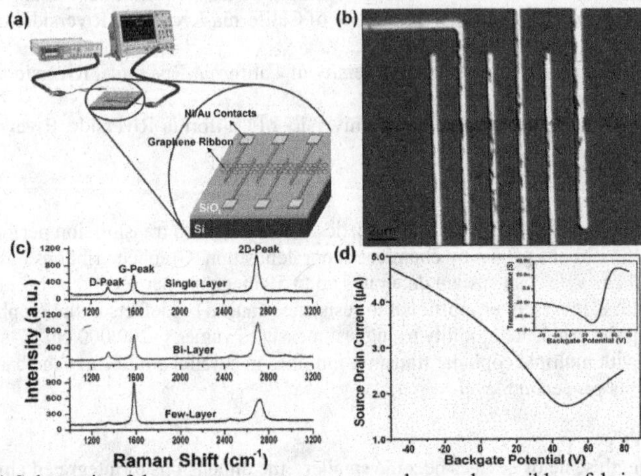

Figure 1. (a) Schematic of the measurement structure and a graphene ribbon fabricated (b) Optical microscope image of device, (c) Raman spectra (d) GR FET source drain current measured according to the backgate potential. The inset shows a plot of transconductance versus back gate potential for the device.

In order to verify the number of graphene layers and their quality, Raman spectroscopy has been used. In Figure 1c, the Raman spectra from single layer, bi-layer and few-layer graphene on Si/SiO_2 substrate are shown, obtained at an excitation wavelength of 514 nm. Positions of the G-Peak and the 2D-Peak in the spectra verify that the films are graphene layers[12] and spectrum variations in the 2D-Peak indicate that the films have several non-uniformly distributed layers which are averaged at 4 layers based on optical transmittance measurements, an average of 89.4% transmittance at 550 nm wavelength.[13] Intensity of the D-Peak indicates the amount of disorder in the graphene layers[14] and it is small in the spectra of our CVD grown graphene layers.

For GRs fabrication, the graphene layers were patterned and etched into stripes using photolithography and Reactive Ion Etch (RIE). Next, back gate field effect transistors with Ni/Au (40 nm/80 nm) electrodes were fabricated.[15] Before characterization, the devices were annealed in Ar/H_2 atmosphere at 400 °C for 30 minutes to remove any photoresist residues on the graphene ribbons. The charge carrier mobilities can be estimated from an analysis of the back gate measurements, done by using Agilent 4155C Semiconductor Parameter Analyzer, shown in Figure 1d by extracting the transconductance, dI_{ds} (drain-source current)/dV_{gs} (backgate potential) characteristics (Figure 1d inset) for back gate potentials above the threshold bias.[16] Hence, the electron and hole mobilities were computed as 2,180 $cm^2V^{-1}s^{-1}$ and 1,453 $cm^2 V^{-1}s^{-1}$, respectively.

For the experimental setup of data transmission, uniformly distributed binary sequences were generated and uploaded to an Agilent 81150A Waveform Generator. Next, the output of the waveform generator was set to generate a signal in which a voltage level of 60 mV represents a one bit string and 10 mv represents a zero bit string. This configuration provides a signal having a 50 mV amplitude and a 10 mV offset. The binary sequence was repeatedly transmitted with a frequency selected according to the data transmission rate. Signals transmitted through the GRIs and test equipment were collected using an Agilent 7000 Series digital oscilloscope for further analysis.

RESULTS AND DISCUSSION

To investigate the digital data transmission performance of the GRs, eye diagram approach is used.[17] The resulting eye diagram for two-symbols per trace measured on a GR which is 1.5 μm in length and 1 μm in width at 50 Megabits per second (Mbps) data rate is given in Figure 2a. Here, the time offset shows the number of symbols and one symbol period is unit interval (u.i.) which is the reciprocal of the data rate.

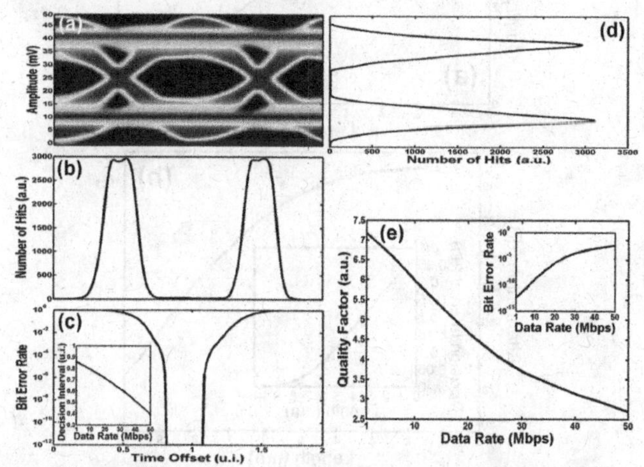

Figure 2. (a) Eye diagram (b) Jitter histogram (c) BER bathtub plots (d) Amplitude histogram (e) Quality factor versus the data rate. Inset shows the BER value vs data rate

One of the most important performance parameters of digital data transmission is the BER. Errors or misread bits can be caused by shifts in time and/or amplitude. First, the probability density function (PDF), $f_{te}(t)$, of the timing errors was calculated at the crossing points of the eye diagram along the time-axis by using histograms.[18] The resulting histogram given in Figure 2b is known as the jitter histogram. For BER determination, the bathtub plot which shows the eye opening on a certain BER value can be used. Bathtub plots can be extracted by calculating the cumulative distribution function (CDF) of the jitter histogram. The CDF integration for a bathtub plot is given in Equations 1 and 2:

$$BER_{head}(t) = \int_{t}^{\infty} f_{te}\left(t'\right)dt' \tag{1}$$

$$BER_{trail}(t) = \int_{-\infty}^{t} f_{te}\left(t'\right)dt' \tag{2}$$

In serial data transmission, timing error analysis[27] is done at a BER level of 10^{-12}. The opening at that level can be assumed as the optimum decision interval.[19] In the inset of Figure 2c, the decision interval according to the data rate is given and it shows that, as the data rate increases, the optimum decision interval linearly decreases.

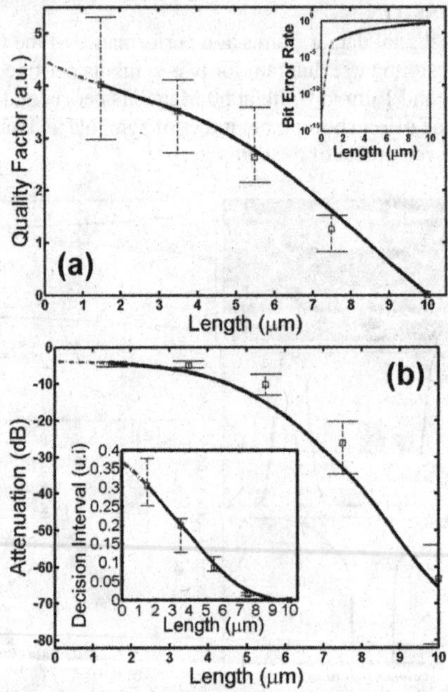

Figure 3. (a) Averaged quality versus length of GRs Inset shows the calculated BER values versus the length of the GRs. (b) Averaged attenuation values versus the length of GRs. Inset shows the optimum decision interval versus the length of the GRs.

On the other hand, the shifts in amplitude value cause errors in the optimum decision interval and the actual BER in this interval can be calculated by a Q factor analysis. For this, first the PDF of the amplitude values were calculated at the center of the optimum decision interval, the optimum decision point, along the amplitude axis by using the amplitude histograms (Figure 2d). The Q factor, the signal-to-noise ratio (SNR) at the decision point,[20] can be calculated by using the mean value (μ_i) and the standard deviation (σ_i) of the histograms as given by,[17]

$$Q = \frac{|\mu_1 - \mu_0|}{\sigma_1 + \sigma_0} \tag{3}$$

where, i = 1 indicates the histogram of one bits and i = 0 indicates the histogram of zero bits. The calculated Q factor according to the data rate is given in Figure 2e. The Q factor decreases almost linearly as the data rate increases. The BER at the optimum decision point can be calculated by using,[19]

$$BER = \frac{1}{2}\text{erfc}\left(\frac{Q}{\sqrt{2}}\right) \tag{4}$$

where erfc is the complementary error function.[20] The BER at the optimum decision point according to the data rate is given in the inset of Figure 2e and the BER value for a data rate level of 50 Mbps is approximately equal to 10^{-4}.

To understand the effect of device length on data transmission, measurement and analysis presented above were repeated for the rest of the contacts fabricated on several graphene ribbons having same geometries and sizes with the one given in Figures 1a and 1b. The averaging Q factor values and variation ranges were calculated for different lengths by using the amplitude histograms and Equation 3. The results shown in Figure 3a indicate that the Q factor depends exponentially on the length of the GRs. In the inset of Figure 3a, the BER according to length is given. The BER values were calculated at the optimum decision points by using Equation 4, which increase by several orders of magnitude within the first micron length. The reason for this huge dependency is the decrease in the vertical opening of the eye diagram due to the attenuation of the signal. In general, the signal attenuates exponentially as the length of the GR increases (Figure 3b). This increase in the attenuation makes the Q factor decrease to intolerable values at which the noise levels are higher than the signal itself. The averaging optimum decision interval and the variation ranges according to length given in the inset of Figure 3b was obtained from timing error analysis, which indicates that the eye has a complete closure after 8.5 μm of length due to the excessive amount of timing errors, i.e. the symbols completely merge.

CONCLUSIONS

In summary, although the data transmission performance of CVD grown few-layer GRs can not compete with that of Cu interconnects (2.5 Gbps over a 20 cm length)[21] at current graphene quality and test structure, the performance is promising for Mbps range applications. Improving the graphene layer quality to the level of physically exfoliated graphene may increase the data rate to hundreds of Gbps levels, and achieving the highest mobility levels measured (200,000 $cm^2V^{-1}s^{-1}$)[6] for suspended graphene layers by designing suspended GRs, may carry the data rates beyond terabits per second (Tbps) levels.

ACKNOWLEDGMENTS

The authors gratefully acknowledge financial support by the CMMI Division of the National Science Foundation (Award: 0800680), the UMASS Materials Research Science and Engineering Center (NSF-MRSEC) on Polymers (Award: 0213695), the UMASS Nanoscale Science and Engineering Center (NSF-NSEC) on hierarchical manufacturing (CHM, Award: 0531171), and the University of California-Riverside.

REFERENCES

1. P. K. Bondyopadhyay, Proceedings of the Ieee **86** (1), 78-81 (1998).
2. A. E. Yarimbiyik, H. A. Schafft, R. A. Allen, M. E. Zaghloul and D. L. Blackburn, Microelectronics Reliability **46** (7), 1050-1057 (2006).
3. Q. Shao, G. Liu, D. Teweldebrhan and A. A. Balandin, Applied Physics Letters **92** (20), 202108 (2008).
4. K. S. Novoselov, A. K. Geim, S. V. Morozov, D. Jiang, Y. Zhang, S. V. Dubonos, I. V. Grigorieva and A. A. Firsov, Science **306** (5296), 666-669 (2004).
5. K. S. Novoselov, A. K. Geim, S. V. Morozov, D. Jiang, M. I. Katsnelson, I. V. Grigorieva, S. V. Dubonos and A. A. Firsov, Nature **438** (7065), 197-200 (2005).
6. K. I. Bolotin, K. J. Sikes, Z. Jiang, M. Klima, G. Fudenberg, J. Hone, P. Kim and H. L. Stormer, Solid State Communications **146** (9-10), 351-355 (2008).
7. D. A. B. Miller and H. M. Ozaktas, Journal of Parallel and Distributed Computing **41** (1), 42-52 (1997).
8. Personic.Sd, Bell System Technical Journal **52** (6), 843-874 (1973).
9. I. Shake, F. Takara and S. Kawanishi, IEEE Photonics Technology Letters **15** (4), 620-622 (2003).
10. A. Reina, X. T. Jia, J. Ho, D. Nezich, H. B. Son, V. Bulovic, M. S. Dresselhaus and J. Kong, Nano Letters **9** (8), 3087-3087 (2009).
11. A. Reina, H. B. Son, L. Y. Jiao, B. Fan, M. S. Dresselhaus, Z. F. Liu and J. Kong, Journal of Physical Chemistry C **112** (46), 17741-17744 (2008).
12. I. Calizo, A. A. Balandin, W. Bao, F. Miao and C. N. Lau, Nano Letters **7** (9), 2645-2649 (2007).
13. R. R. Nair, P. Blake, A. N. Grigorenko, K. S. Novoselov, T. J. Booth, T. Stauber, N. M. R. Peres and A. K. Geim, Science **320** (5881), 1308-1308 (2008).
14. A. C. Ferrari, J. C. Meyer, V. Scardaci, C. Casiraghi, M. Lazzeri, F. Mauri, S. Piscanec, D. Jiang, K. S. Novoselov, S. Roth and A. K. Geim, Physical Review Letters **97** (18), 187401 (2006).
15. I. Suarez-Martinez, A. Felten, J. J. Pireaux, C. Bittencourt and C. P. Ewels, Journal of Nanoscience and Nanotechnology **9** (10), 6171-6175(6175) (2009).
16. B. G. Streetman, *Solid State Electronic Devices*, 4 ed. (Prentice Hall, New York, 1995).
17. I. Shake, H. Takara and S. Kawanishi, Journal of Lightwave Technology **22** (5), 1296-1302 (2004).
18. C.-K. Ong, D. Hong, K.-T. T. Cheng and L.-C. Wang, presented at the Proceedings of the Design, Automation and Test in Europe Conference and Exhibition Paris, (2004).
19. J. D. Downie, Journal of Lightwave Technology **23** (6), 2031-2038 (2005).
20. N. S. Bergano, F. W. Kerfoot and C. R. Davidson, IEEE Photonics Technology Letters **5** (3), 304-306 (1993).
21. S. Jaernin, S. Chung-Seok, A. Chellappa, M. Brooke, A. Chattejce and N. M. Jokerst, presented at the Electronic Components and Technology Conference, 2003. Proceedings. 53rd, (2003).

Topological Insulators and Quasi-2D Materials

Mater. Res. Soc. Symp. Proc. Vol. 1344 © 2011 Materials Research Society
DOI: 10.1557/opl.2011.1349

Low-Frequency Noise in "Graphene-Like" Exfoliated Thin Films of Topological Insulators

M. Z. Hossain[1], S. L. Rumyantsev[2,3], K. M. F. Shahil[1], D. Teweldebrhan[1], M. Shur[2] and A. A. Balandin[1,*]

[1]Nano-Device Laboratory, Department of Electrical Engineering and Materials Science and Engineering Program, University of California, Riverside, California 92521 USA
[2]Department of Electrical, Computer and Systems Engineering and Center for Integrated Electronics, Rensselaer Polytechnic Institute, Troy, New York 12180 USA
[3]Ioffe Institute, Russian Academy of Sciences, St. Petersburg, 194021 Russia

ABSTRACT

We report results of the study of the low-frequency noise in thin films of bismuth selenide topological insulators, which were mechanically exfoliated from bulk crystals via "graphene-like" procedures. From the resistance dependence on the film thickness, it was established that the surface conduction contributions to electron transport were dominant. It was found that the current fluctuations have the noise spectral density $S_I \propto 1/f$ (where f is the frequency) for the frequency range up to 10 kHz. The obtained noise data are important for transport experiments with topological insulators and for any proposed device applications of these materials.

INTRODUCTION

Bismuth selenide (Bi_2Se_3) is a representative of a newly discovered class of materials referred to as topological insulators (TI). These materials exhibit quantum-Hall-like behavior in the absence of a magnetic field. TIs reveal conducting surface states that are protected against scattering by the time-reversal symmetry [1-6]. The proposed applications of TIs include quantum computing, spintronics, magnetic memory and low-energy dissipation electronics. Topological insulators have also shown promise for thermoelectric applications [7, 8]. Bismuth selenide (Bi_2Se_3) and other related materials such as Bi_2Te_3 and Sb_2Te_3 have recently been identified as three-dimensional (3D) topological insulators with a single Dirac cone at the surface [4-6]. Among 3D TIs, Bi_2Se_3 serves as a reference material owing to its relatively simple band structure and large bulk band gap of approximately 0.3 eV [4].

The identification between the volume and the surface conduction in TIs brings up analogies with such a ubiquitous phenomenon as $1/f$ noise, which could be linked to either volume or surface conduction or both. The low-frequency fluctuations in electrical current have a well defined $1/f$ spectral density for the frequency f below ~10–30 kHz [9]. From the application point of view, signal-to-noise ratio (SNR) of TI devices should be as large as possible. The unavoidable up-conversion of $1/f$ noise limits the phase noise of the oscillators and is a major concern for RF and analog/mixed-signal circuits. Moreover, LFN measurement offers a strong characterization method for materials quality and reliability evolution of devices. The studies of low-frequency $1/f$ noise in materials and thin films which have been identified as topological insulators are of critically importance for identification of volume versus surface transport. Here we will focus on $1/f$ electronic noise of exfoliated Bi_2Se_3 thin-film devices.

EXPERIMENTAL APPROACH

The bulk crystal structure of Bi_2Se_3 is trigonal with lattice constant of a= 0.984 nm and α= 25° [10]. Its structure is visualized in terms of a layered structure with each layer having a thickness of ~1 nm referred to as a quintuple layer (QL). Each quintuple layer consists of a five atomic planes arranged in the sequence of …..Se(1)–Bi–Se(2)–Bi–Se(1)… Two quintuple layers are weakly bonded to each other by van der Waals forces. The crystal has inversion symmetry with respect to the center Se(2) atoms. The weak bonding between quintuple layers (Se(1)-Se(1)) allows one to obtain quasi two-dimensional atomic quintuple films via the "graphene-like" mechanical exfoliation. The details of the process, which we developed for a similar Bi_2Te_3 crystal, were reported elsewhere [7, 11]. The exfoliation from the bulk samples allowed us to obtain crystalline films with relatively low defects.

The obtained films were characterized using optical and scanning electron microscopy, and subjected to micro-Raman inspection [12] in order to identify the quality and crystallinity of the films. We intentionally selected rather thick films, with the thickness ranging from ~50 to 170 nm, in order to avoid hybridization of the top and bottom electron surface states. The typical Bi_2Se_3 films were n-type with the carrier density ~10^{18} cm^{-3}. The mobility of such film is on the order of ~200 $cm^2V^{-1}s^{-1}$ at room temperature [13]. After examining the crystallinity of Bi_2Se_3 flake, we fabricated devices using the electron beam lithography, evaporation and lift-off process. All the fabricated devices showed linear current-voltage characteristics in the low-bias regime and low contact resistance compared to the channel resistance. We carried out the LFN measurements using a spectrum analyzer (SR 770) at ambient conditions following the standard measurement protocols [14, 15]. Drain-source current fluctuations were recorded, and analyzed by dynamic signal analyzer after being amplified by low-noise pre-amplifier. The whole setup was kept inside the metal shielding box to reduce the effects of environmental noises and electromagnetic fields.

RESULTS AND DISCUSSIONS

Measuring the resistance R of the device channels for different thicknesses varying from ~50 nm to ~150 nm we established that R does not scale with the thickness as one would expect from Ohm's law. The contact resistance was determined to be much less than that of the device channel. These two facts suggest that the contribution from the surface to the electron transport is dominant or, at least, significant. This means that the electron transport in our devices was close to the surface TI regime.

Figure 1 shows the frequency dependence of the measured noise current spectral density for Ti/Au contacted Bi_2Se_3 thin film devices. One can see that the measured excess noise is close to $1/f$ type for the frequency range up to 10^4 Hz. The spectra are free from the generation-recombination bulges. The latter suggests that the samples are of relatively high quality and no single trap dominates the noise. We fitted the experimental data with the functional dependence of $1/f^{\beta}$ with β value close to ~1.10. As expected for $1/f$ noise, we found the noise spectral density S_I to increase with the increase of source-drain voltage [14].

We repeated the noise measurement for a number of devices with various channel resistances and introduced the amplitude of the current noise as $S_I = AI^2/f$ [16]. Here A is the relative noise amplitude, which is a parameter convenient for comparing $1/f$ noise levels in

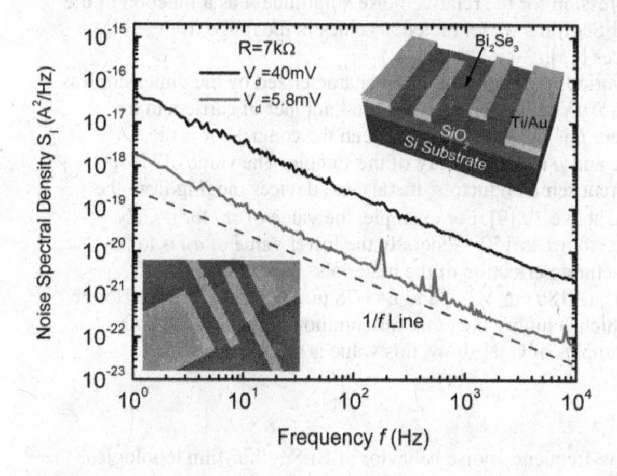

Figure 1. Noise current spectral density for a typical Bi_2Se_3 thin film TI devices. The resistance of the representative device is 7 kΩ. Insets show the schematic representation and SEM image of the fabricated four-contact topological insulator device.

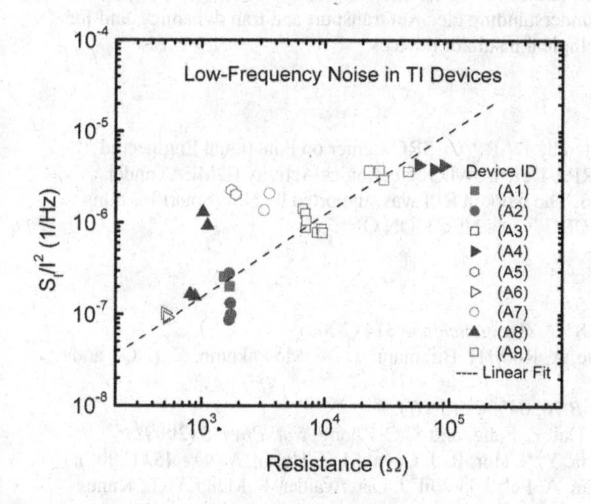

Figure 2. Relative noise amplitude S_I/I^2 at f=1Hz for several Bi_2Se_3 thin-film topological insulator device as a function of the device resistance.

various materials and devices. Figure 2 shows the normalized noise power (S_I/I^2) at f=1 Hz as a function of the device resistance. We found that Bi_2Se_3 devices with a higher resistance had the higher normalized noise spectral density. The dash line shown in Figure 2 is the least-squares power-law fit, which yields the expression for the relative noise amplitude A as a function of the resistance R: $A=3\times10^{-10}R^{-0.9}$. For Bi_2Se_3 films we extracted A values in the range from $\sim10^{-7}$ to $\sim10^{-5}$ for the examined Bi_2Se_3 devices [15].

The 1/f noise properties of various materials are often characterized by the dimensionless empirical Hooge parameter, $\alpha_H=(S_I/I^2)fN$ [17]. Here N is the total number of carriers in the sample estimated as $N=L^2/Rq\mu$ where L is the separation between the contacts, R is the resistance, q is the elemental charge and μ is the mobility of the sample. The value of Hooge parameter can be different for different semiconductors, metals and devices ranging from the very low, below $\sim10^{-8}$, to very high, above 10 [9]. For example, the value of α_H for p-GaN resistor reported in Ref. [18] ranged from 1 to 150. Generally the lower value of α_H is to be expected for the higher level of structural perfection of the materials. Using the value of R=526.87 Ω (from dc measurement), $\mu\approx180$ cm^{2}V^{-1}s^{-1} and L=1.78 μm, we estimated the Hooge parameter for our sample is ~0.2 which is higher than that for common semiconductors or metals. At the same time, as the examples of GaN show, this value is not too unusual.

CONCLUSIONS

We have investigated the low-frequency noise behavior of Bi_2Se_3 thin-film topological insulator devices. We found that the current fluctuations in this type of materials had the pure 1/f spectral density for the frequency below 10 kHz with no clear evidence of generation recombination bulges. From the measured dependence of resistance on the film thickness we concluded that the surface contribution to electron conduction was substantial in our samples. The obtained results can be used for understanding electron transport and trap dynamics, and for reducing low-frequency noise in topological insulator devices.

ACKNOWLEDGEMENTS

The work at UCR was supported by DARPA – SRC Center on Functional Engineered Nano Architectures (FENA) and DARPA Defense Microelectronics Activity (DMEA) under agreement number H94003-10-2-1003. The work at RPI was supported by NSF Smart Lighting Engineering Research Center and I/UCRC 'CON NECTION ONE'.

REFERENCES

[1] A. B. Bernevig, T. L. Huges, and S. C. Zhang, *Science* **314** (2006).
[2] M. Konig, S. Wiedmann, C. Brune, A. Roth, H. Buhmann, L. W. Molenkamp, X. L. Qi, and S. C. Zhang, *Science* **318** (2007).
[3] L. Fu, and C. L. Kane, *Phys. Rev. B* **76**, 045302 (2007).
[4] H. Zhang, C. X. Liu, X. L. Qi, X. Dai, Z. Fang, and S. C. Zhang, *Nat. Phys.* **5** (2009).
[5] D. Hsieh, D. Qian, L. Wray, Y. Xia, Y. S. Hor, R. J. Cava, M. Z. Hasan, *Nature* **452** (2008).
[6] D. Hsieh, Y. Xia, L. Wray, D. Qian, A. Pal, J. H. Dil, J. Osterwalder, F. Meier, C. L. Kane, and G. Bihlmayer, *Science* 323 (2009).
[7] D. Teweldebrhan, V. Goyal, and A. A. Balandin, *Nano Lett.* 10 (2010)

[8] F. Zahid, and R. Lake, *Appl. Phys. Lett.* **97**, 212102 (2010).
[9] A. A. Balandin, "Noise and Fluctuations Control in Electronic Devices," ASP (2002).
[10] R. W. G Wyckoff, "Crystal Structures," (Krieger, Melbourne, FL, Vol. 2, 1986).
[11] D. Teweldebrhan, V. Goyal, M. Rahman, and A. A. Balandin, *Appl. Phys. Lett.* **96**, 053107 (2010).
[12] K. M. F. Shahil, M. Z. Hossain, D. Teweldebrhan, and A. A. Balandin, *Appl. Phys. Lett.* **96**, 153103 (2010).
[13] H. Steinberg, D. R. Gardner, Y. S. Lee, and P. J. Herrero, *Nano Lett.* **10** (2010).
[14] M. Z. Hossain, S. L. Rumyantsev, D. Teweldebrhan, K. M. F. Shahil, M. Shur and A. A. Balandin, *Physica Status Solidi (A)* **1** (2011).
[15] M. Z. Hossain, S. L. Rumyantsev, D. Teweldebrhan, K. M. F. Shahil, M. Shur and A. A. Balandin, *ACS Nano*, DOI: 10.1021/nn102861d.
[16] P. G. Collins, M. S. Fuhrer, and A. Zettl, *Appl. Phys. Lett.* **76** (2000).
[17] F. N. Hooge, and L. K. J. Vandamme, *Phys. Lett. A* **66** (1978).
[18] A. K. Rice, K. J. Malloy, and K. J. Bulk, *J. Appl. Phys.*87 (2000).

Mater. Res. Soc. Symp. Proc. Vol. 1344 © 2011 Materials Research Society
DOI: 10.1557/opl.2011.1361

Pseudo-Superlattices of Bi_2Te_3 Topological Insulator Films with Enhanced Thermoelectric Performance

V. Goyal, D Teweldebrhan and A.A. Balandin

Nano-Device Laboratory, Department of Electrical Engineering and Materials Science and Engineering Program, Bourns College of Engineering, University of California, Riverside, California 92521 USA.

ABSTRACT

It was recently suggested theoretically that atomically thin films of Bi_2Te_3 topological insulators have strongly enhanced thermoelectric figure of merit. We used the "graphene-like" exfoliation process to obtain Bi_2Te_3 thin films. The films were stacked and subjected to thermal treatment to fabricate *pseudo-superlattices* of single crystal Bi_2Te_3 films. Thermal conductivity of these structures was measured by the "hot disk" and "laser flash" techniques. The room temperature in-plane and cross-plane thermal conductivity of the stacks decreased by a factor of ~2.4 and 3.5 respectively as compared to that of bulk. The strong decrease of thermal conductivity with preserved electrical properties translates to ~140-250% increase in the thermoelectric figure if merit. It is expected that the film thinning to few-quintuples, and tuning of the Fermi level can lead to the topological insulator surface transport regime with the theoretically predicted extraordinary thermoelectric efficiency.

INTRODUCTION

Bismuth Telluride (Bi_2Te_3) and its alloys are known as the best thermoelectric (TE) materials [1] with room temperature thermoelectric figures of merit $ZT=S^2\sigma T/(K_e+K_l) \sim 1$, where S is the Seebeck coefficient, σ is electrical conductivity, T is the absolute temperature, K_e and K_l are the electron and phonon (lattice) contributions to the thermal conductivity. It has been suggested by theoretical predictions that a drastic enhancement in ZT can be achieved in low-dimensional structures either from confinement-induced increased electron density of states near the Fermi level E_F [2-3] or reduction of K_l due to the phonon – boundary scattering or modification of the phonon spectrum [4].

Most recently, it was shown that the Bi_2Te_3 related materials such as Bi_2Se_3 and Sb_2Te_3 are topological insulators (TIs). TIs are materials with a bulk insulating gap and conducting surface states that are topologically protected against scattering by the time-reversal symmetry [5-6]. It was shown theoretically that ZT can be strongly enhanced in Bi_2Te_3 thin-film TIs provided that the Fermi level is tuned to ensure the surface transport regime and the films are thin enough to open a gap in the "Dirac cone" dispersion on the surface [7]. At the same time, thermoelectric applications require a sufficient quantity of material, i.e. "bulk", i.e. the single quintuples would hardly be practical. For this reason, we studied the stacks of the exfoliated films, which were put on top of each other and subjected to thermal treatment.

EXPERIMENTAL DETAILS

The "pseudo-superlattices" were fabricated by the stacking of individual single-crystal Bi_2Te_3 films exfoliated using the "graphene-like" mechanical exfoliation method. We have previously demonstrated that the "graphene-like" procedure can be used to mechanically exfoliate the ultra-thin films of Bi_2Te_3 with the thickness down to a single *quintuple* [8]. These exfoliated thin films with different thicknesses (ranging from few nm to µm) were then transferred to a substrate and mechanically put on each other to form a stack followed by annealing at ~250^{0}C for 30 second to reduce the air gaps between the layers. Different stacked samples with varying thicknesses (up to ~0.5 mm) were fabricated to study thermoelectric properties. Figure 1 shows cross-sectional SEM (Philips XL-30 FEG) image of stacked samples of exfoliated Bi_2Te_3 films revealing that the films of different thicknesses are stacked into a non-periodic "pseudo-superlattice". This non-periodicity in the superlattice structure was intentionally introduced to make advantage of certain benefits for thermoelectric applications owing to flexibility for tuning the phonon transport in non-periodic superlattices [9].

Figure 1: SEM image of the stacked "pseudo-superlattice" of the mechanically exfoliated Bi_2Te_3 films.

The material crystallinity and quality was verified with micro-Raman spectroscopy (Renishaw) used in a backscattering configuration. Figure 2 shows a spectrum of the exfoliated Bi_2Te_3 film (with thickness ~50nm) recorded using 488-nm laser excitation. The thickness was cross-checked with the atomic force microscopy. The spectra was recorded with 1800 lines/mm grating providing the spectral resolution of ~ 1 cm^{-1} (software enhanced resolution is ~0.5 cm^{-1}). The observed spectrum is in line with literature [10]. An additional peak, identified as A_{1u}, also appears for our exfoliated films, which appears as a result of the crystal symmetry breaking in thin films with the thickness below the light penetration depth [11] and is not present in the bulk Bi_2Te_3 crystals.

The measurements of the thermal conductivity, K, on these stacks were performed by two different experimental techniques. The first technique was the transient plane source (TPS) "hot disk" technique, which measures the average in-plane thermal conductivity. The second technique was the optical "laser flash" technique (LFT), which measures the average cross-plane K. In TPS technique [12-13], an electrically insulated sensor is sandwiched between two pieces of the sample under investigation. During measurement, a short electric pulse is passed through the sensor, which generates heat. The sensor acts both as a heat source and a thermometer to determine the temperature rise, ΔT, in the sample as a function of time, which is used to determine the thermal conductivity. Our LFT instrument (Netzsch NanoFlash) was equipped with a xenon flash lamp which heated the sample from one end by light pulses. The temperature rise was determined at the back side of the samples with the nitrogen-cooled InSb IR detector. Using the thermal-wave travel time, we measured α, and determined K from the equation $K=\alpha\rho C_p$, where ρ is the mass density and C_p is the specific heat of the material.

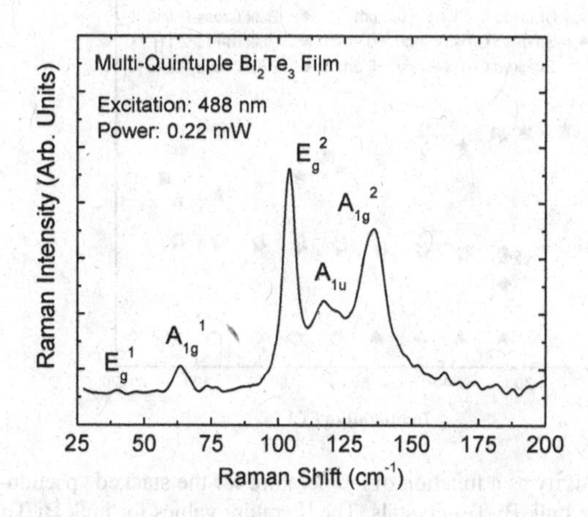

Figure 2: Raman spectrum of the "graphene-like" exfoliated Bi_2Te_3 films. Note the appearance of A_{1u} peak, not Raman active in bulk crystals, due to the crystal symmetry breaking in thin films.

EXPERIMENTAL RESULTS AND DISCUSSION

Figure 3 presents the results of our thermal measurements for three representative "pseudo-superlattices" with thicknesses ~0.4 mm, ~0.3 mm and with ~0.1 mm [14]. A comparison with K values for Bi_2Te_3 from literature [15] attests to the accuracy of our measurements. A strong decrease in both the in-plane as well as cross-plane thermal conductivity of the "pseudo-superlattices" was observed as compared to the bulk. The RT in-plane (cross-plane) K value of the stacks is ~0.7 W/mK (~0.14W/mK), which is a reduction by a factor of ~2.4 (~3.5) from the bulk value of ~1.7 W/mK (~0.5W/mK). The "pseudo-superlattice" K is only weakly dependent on temperature which can be attributed to the thermal transport limited by the phonon – boundary scattering [4, 16]. The thermal conduction in our samples was mostly by the acoustic phonons; the electron contribution was estimated to be ~10% as calculated from the Wiedemann-Franz law. Another interesting observation is that the K values obtained for the stacks with different thicknesses are nearly the same indicating that the thermal conductivity was limited by the phonon scattering at the interfaces between the individual Bi_2Te_3 layers rather than by the scattering on the outside boundaries of the samples. The overall decrease of K in our stacks is exceptional. It is on the higher end of that reported for Bi_2Te_3 nanoparticles [16], alloy films [17] and highly-textured materials [18]. The cross-plane *K* in our "pseudo-superlattices" approaches the theoretical *minimum* value predicted for the disordered crystals [19].

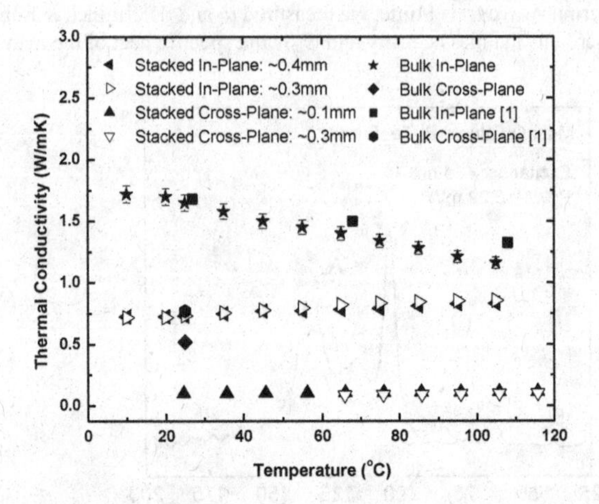

Figure 3: Thermal conductivity as a function of temperature for the stacked "pseudo-superlattices" and reference bulk Bi_2Te_3 crystals. The literature values for bulk Bi_2Te_3 are also shown for comparison.

The Seebeck measurements (MMR SB100) gave *S* values in the range of ~231-247 μV/K. The *I-V* characteristics of the "pseudo-superlattices" were studied using probe station (Signatone) at ambient conditions. The electrical resistivity was measured on the order of ~10^{-4} Ωm, which is close to the optimum for the thermoelectric applications. The strong decrease in the thermal conductivity with preserved electrical properties translates to ~140-250% increase in *ZT* at RT. The estimated ZT enhancement in stacked samples is achieved entirely via reduction in the thermal conductivity. This enhancement can further be increased by thinning of the films and increasing their crystal quality, and gating (for achieving the pure surface transport) is expected to result in additional strong *ZT* increase predicted theoretically [7].

CONCLUSIONS

In conclusions, we studied thermoelectric properties of "pseudo-superlattices" prepared by staking of the "graphene-like" mechanically exfoliated single-crystal Bi_2Te_3 films. We experimentally showed that *ZT* in such structures can be substantially increased via reduction of the in-plane and cross-plane thermal conductivity. It is an important observation, since Bi_2Te_3 were shown to be topological insulators. Eventually, it may become possible to achieve the pure surface transport regime through the Dirac surface states, topologically protected against scattering, and achieve the theoretically predicted strong enhancement of *ZT* over a wide temperature range.

ACKNOWLEDGMENTS

The authors acknowledge the support from SRC – DARPA through the FCRP Center on Functional Engineered Nano Architectonics (FENA) and DARPA Defense Microelectronics Activity (DMEA) under agreement number H94003-10-2-1003.

REFERENCES

1. H. J. Goldsmid, Thermoelectric Refrigeration (Plenum, New York, 1964); D. M. Rowe, CRC Book on Thermoelectrics (CRC Press, 1995).
2. M. S. Dresselhaus, G. Dresselhaus, X. Sun, Z. Zhang, S. B. Cronin and T. Koga, *Physics of the Solid State* **41**, 679 (1999).
3. L. D. Hicks, M. S. Dresselhaus, *Phys. Rev. B.* **47**, 12727 (1993).
4. A. Balandin and K.L. Wang, *Phys. Rev. B.* **58**, 1544 (1998); A. Balandin and K.L. Wang, *J. Appl. Phys.* **84**, 6149 (1998).
5. M.Z. Hasan and C.L. Kane, *Rev. Mod. Phys.* 82, 3045 (2010); X.-L. Qi and S.-C. Zhang, arXiv: 1008.2026.
6. J. Moore, *Nature Phys.* **5**, 378 (2009).
7. P. Ghaemi, R.S.K. Mong and J.E. Moore, *Phys. Rev. Lett.* **105**, 166603 (2010); F. Zahid and R. Lake, *Appl. Phys. Lett.*, **97**, 212102 (2010).
8. D. Teweldebrhan, V. Goyal and A.A. Balandin, *Nano Lett.* **10**, 1209 (2010); D. Teweldebrhan, V.Goyal, M. Rahman, and A. A. Balandin, *Appl. Phys. Lett.* **96**, 053107 (2010).
9. S.A. Barnett and M. Shinn, *Annu. Rev. Mater. Sci.* **24**, 481 (1994); S. Tamura and F. Nori, *Phys. Rev. B.* **41**, 7941 (1990).
10. W. Kullmann, J. Geurts, W. Richter, N. Lehner, H. Rauh, U. Steigenberger, G. Eichhorn and R. Geick, *Phys. Stat. Sol. (b)* **125**, 131 (1984); W. Richter, H. Kohler and C. R. Becker, *Phys. Stat. Sol. (b)* **84**, 619 (1977).
11. K.M.F. Shahil, M.Z. Hossain, D. Teweldebrhan and A.A. Balandin, *Appl. Phys. Lett.* **96**, 153103 (2010)
12. S. E. Gustafsson, *Rev. Sci. Instrum.* **62**, *797* (1991).
13. S. Ghosh, D. Teweldebrhan, J. R. Morales, J. E. Garay, and A. A. Balandin, *J. Appl. Phys.* **106**, 113507 (2009); R. Ikkawi, N. Amos, A. Lavrenov, A. Krichevsky, D. Teweldebrhan, S. Ghosh, A.A. Balandin, D. Litvinov, S. Khizroev, *J. Nanoelectron. Optoelectron.* **3**, 44 (2008).
14. V. Goyal, D. Teweldebrhan, and A. A. Balandin, *Appl. Phys. Lett.* **97**, 133117 (2010).
15. C. B. Satterthwaite and R. W. Ure, Jr., *Phys. Rev.* **108**, 1164 (1957).
16. M.R. Dirmyer, J. Martin, G.S. Nolas, A. Sen, J.V. Badding, *Small* **5**, 933 (2009).
17. C. Chiritescu, C. Mortensen, D.G. Cahill, D. Johnson and P. Zschack, *J. Appl. Phys.* **106**, 073503 (2009).
18. O. Ben-Yehuda, R. Shuker, Y. Gelbstein, Z. Dashebsky and M.P. Dariel, *J. Appl. Phys.* **101**, 113707 (2007).
19. D. Cahill, S. Watson, R. Pohl, *Phys. Rev. B.* **46**, 6131 (1992).

Mater. Res. Soc. Symp. Proc. Vol. 1344 © 2011 Materials Research Society
DOI: 10.1557/opl.2011.1359

" Graphene-Like" Exfoliation of Quasi-2D Crystals of Titanium Ditelluride: A New Route to Charge Density Wave Materials

Javed M. Khan, Desalegne Teweldebrhan, Craig M. Nolen and Alexander. A. Balandin

Nano-Device Laboratory, Department of Electrical Engineering and Material Science and Engineering Program, Bourns College of Engineering, University of California – Riverside, California 92521, USA

ABSTRACT

We used a "graphene-like" mechanical exfoliation to obtain atomically thin films of $TiTe_2$. The building blocks of titanium ditelluride are atomic tri-layers separated by the van der Waals gaps. The exfoliation procedure allows one to obtain the few-atom-thick films with strong confinement of charge carriers and phonons. We have verified the crystallinity of the exfoliated films and fabricated the back-gated field-effect devices. The current – voltage characteristics of the $TiTe_2$ devices revealed strong non-linearity, which suggests the charge-density wave effects. The obtained results are important for the proposed application of $TiTe_2$ for the charge-density wave devices and thermoelectric energy conversion.

INTRODUCTION

The mechanical exfoliation of graphene (1, 2) stimulated interest to the atomically thin films of other materials. Graphene revealed many interesting properties such as extremely high electron mobility (1, 2) and superior thermal conductivity (3, 4). Recently, we extended the "graphene-like" exfoliation approach to obtain quintuples – atomic five-layers – of Bi_2Te_3 and related materials (5-7). The atomically thin films of Bi_2Te_3 family are important for thermoelectric and topological insulator applications. It was found that the properties of Bi-Te quintuples are substantially different from those of bulk crystals (6-7).

The layered transition-metal-dichalcogenides and related materials constitute another interesting group of materials for exfoliation. These materials reveal the charge density wave (CDW) effects (8-18) and were also investigated for applications in low temperature thermoelectrics. The presence of the van der Waals gap in the unit cell allows one to disassemble $TiTe_2$ into its building blocks - three mono-atomic sheets Te – Ti – Te or its derivatives. The exfoliation procedure allows one to obtain the few-atom-thick single-crystal films with strong confinement of charge carriers and phonons. The spatial confinement effects can be used to optimize the obtained materials for various proposed applications.

EXPERIMENTAL DETAILS

Using the standard mechanical exfoliation procedure (1-6) we obtained the atomically thin films of the individual tri-layers of Te - Ti – Te or few-tri-layer films (see Figure 1). The tri-layer thickness is ~0.27 nm. We carried out detailed micro-Raman spectroscopy investigation of the exfoliated material. The exfoliated films of $TiTe_2$ were initially identified via the optical contrast method, scanning electron microscopy (SEM), and transmission electron microscopy (TEM).

The exfoliated atomically-thin films were transferred on top of the 300-nm oxide layer in Si/SiO_2 substrates. The micro-Raman spectroscopic inspection (InVia, Renishaw) was preformed in a backscattering configuration using laser excitation in the visible light (λ = 488 nm). The spectra were collected with the 100X objective and 1800 line/mm grating. To avoid local heating and melting all spectra were collected at very low excitation power levels ($P < 0.5$ mW). The optimum excitation power (measured at the laser) was 0.35 mW for our ultrathin $TiTe_2$ films. A more detailed description of the micro-Raman spectroscopy of this type of films was reported elsewhere (19). The diffraction patterns of the crystalline structures of the layers were studied using transmission electron microscopy (TEM). The sample preparation for TEM (FEI-PHILIPS CM300) inspection was carried out in the same method as Si/SiO_2 samples preparations. The exfoliated atomically-thin films were transferred on top the a-carbon TEM grids.

RESULTS AND DISCUSSION

We used the high-resolution TEM to examine the crystalline structure of the obtained films. In Figure 1, a TEM image demonstrates the layered structure of the $TiTe_2$ films while the electron diffraction pattern indicates crystallinity.

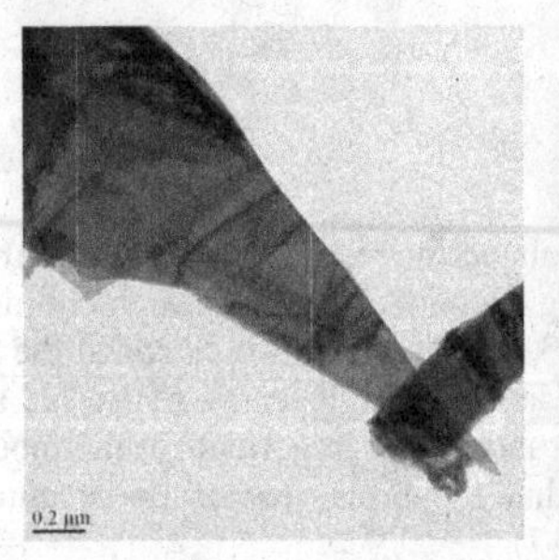

Figure 1: a) TEM micrograph of the "graphene-like" exfoliated films of $TiTe_2$; b) electron diffraction pattern of the $TiTe_2$ thin film indicating its single-crystal nature. The exfoliated films can allow one to obtain the regime of strong quantum confinement for both electrons and phonons.

One of our motivations for the study of the current voltage characteristics of the $TiTe_2$ thin crystalline films was to obtain the signatures of CDW effects in this material at room temperature (RT). The spatial confinement effects may affect the transition temperature to CDW regime. The CDW conductors are usually characterized by non-linear current-voltage (I-V) curves and electrical anisotropy. We were able to fabricate the back-gated field-effect transistor (FET) type devices with $TiTe_2$ films as the conduction channels. Figure 2 shows the measured source-drain current as a function of the source-drain voltage. One can see that the current strongly increases when the voltage reaches a certain threshold voltage V_T. This behavior is similar to the one observed in CDW materials. The on-and-off state switching is not as pronounced as in some other CDW materials possibly because the transition temperature in $TiTe_2$ in bulk form is below RT. The back-gating function in our $TiTe_2$ devices was not as good as in graphene FETs but better than in devices fabricated with exfoliated Bi_2Te_3 films.

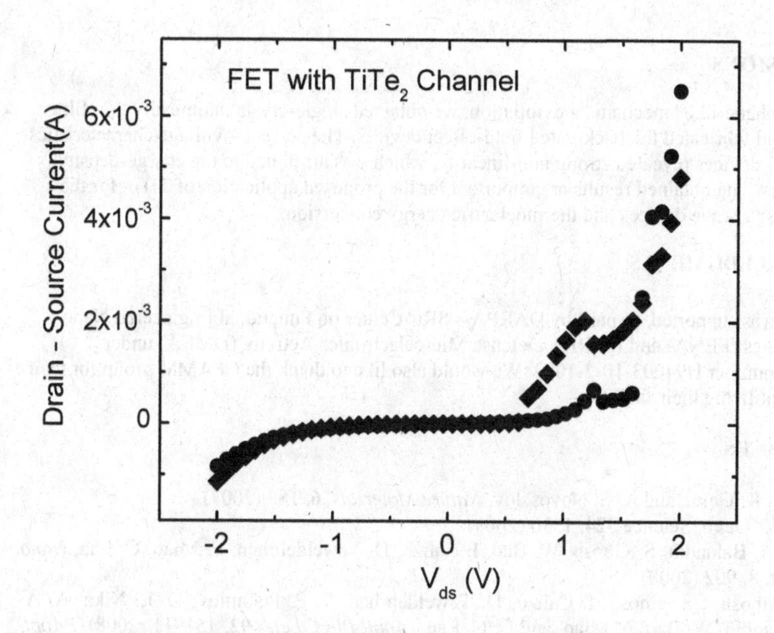

Figure 2: Electrical characteristics of FET type of device with the channel made from the crystalline thin film of $TiTe_2$. The presence of the voltage threshold for the on-set of the electrical current suggests the CDW behavior. In this case, the measured current should be interpreted as the collective CDW current.

In the Table I we summarized the threshold voltage values for different gate biases. One should note here that the gate bias is high due to the thick back oxide used in the device. There is a clear shift in the I-V curve and V_T with the gate bias. It can be related to the changes in the strain distribution resulting from the high gate bias. These shifts in the I-V curves are known as CDW phase slips (17, 20-22). Positive gate voltages greater than 30 volts slideS the I-V curve and V_T to the right. Applied negative gate voltages greater than -50 volts produce a slide the I-V curve and V_T to the left. Interestingly, data suggest CDW V_T can be controlled through gating effect.

Table I. Back-Gated I-V Curves

Back-Gate Voltage V_G (V)	V_{DS} (mV)	I_{DS} (mA)	V_T (mV)
50 < Vg < 30	653	0.772	755
10 < Vg < -30	483	0.719	605
Vg = -50	306	0.754	387

CONCLUSIONS

Using "graphene-like" mechanical exfoliation, we obtained single-crystal atomically thin films of $TiTe_2$, and fabricated the back-gated field-effect devices. The current – voltage characteristics of the $TiTe_2$ devices revealed strong non-linearity, which was attributed to the charge-density wave effects. The obtained results are important for the proposed application of $TiTe_2$ for the charge-density wave devices and thermoelectric energy conversion.

ACKNOWLEDGMENTS

This work was supported, in part, by DARPA – SRC Center on Functional Engineered Nano Architectonics (FENA) and DARPA Defense Microelectronics Activity (DMEA) under agreement number H94003-10-2-1003. We would also like to thank the CFAMM group for their support in utilizing their facilities.

REFERENCES

1. 1 A. K. Geim, and K. S. Novoselov, *Nature Materials*, **6**, 183 (2007).
2. A. K. Geim, Science **324**, 1530 (2009).
3. A. A. Balandin, S. Ghosh, W. Bao, I. Calizo, D. Teweldebrhan, F. Miao, C. Lau, *Nano Lett.* **8**, 902 (2008).
4. S. Ghosh, , S. Ghosh, I. Calizo, D. Teweldebrhan, E. P. Pokatilov, D. L. Nika, A. A. Balandin, W. Bao, F. Miao, and C. N. Lau,' *Appl. Phys. Lett.*, **92**, 151911 (2008).., *Appl. Phys. Lett.*, **92**, 151911 (2008)
5. D. Teweldebrhan, V. Goyal, M. Rahman, and A. A. Balandin, *Appl. Phys. Lett.*, **96**, 053107 (2010).
6. D. Teweldebrhan, V. Goyal, and A. A. Balandin, *Nano Lett.* **9**, 12 (2009).
7. K. M. F. Shahil, M. Z. Hossain, D. Teweldebrhan, and A. A. Balandin, *Appl. Phys. Lett.*, **96**, 153103 (2010).
8. F. Clerc, C. Battaglia, M. Bovet, L. Despont, C. Monney, H. Cercellier, M. Garnier, and P. Aebi, *Physical Review B*, **74**, 155114, (2006)
9. J. A. Wilson, F. J. Di; and S. Mahajan, *Advances in Physics*, **50**: 8, 1171-1248, (2001)

10. H. Cercellier, C. Monney, F. Clerc, C. Battaglia, L. Despont, M. G. Garnier, H. Beck, and P. Aebi, *Phys. Rev. Lett.* **99**, 146403 (2007)
11. R. E. Thorne, *Physics Today*, (1996)
12. H. S. J. van der Zant, N Markovic, E Slot, *Quantum dot and wells, mesocopic networks*, (2001)
13. J. F. Zhao, H.W. Ou, G. Wu, B.P. Xie, Y. Zhang, D. W. Shen, J. Wei, L. X. Yang, J. K. Dong, M. Arita, H. Namatame, M. Taniguchi, X. H. Chen, and D. L. Feng., arXiv:cond-mat/0612091v3, (2008)
14. J. F. Zhao, H.W. Ou, G. Wu, B.P. Xie, Y. Zhang, D. W. Shen, J. Wei, L. X. Yang, J. K. Dong, M. Arita, H. Namatame, M. Taniguchi, X. H. Chen, and D. L. Feng., arXiv:cond-mat/0612091v3, (2008)
15. A. V. Postnikov, *Com. Mat. Sci.*, (2003)
16. S. N. Patel, and A. A. Balchin, *J. Mat. Sci. Lett.*, **4**, (1985)
17. A. Cingolani, M. Lugara, and G. Scamarcio, *Sol. St. Com.*, **62**, (1987)
18. G. Lucovsky, and R. M. White, *Phy Rev B*, **8,** (1973)
19. J. Khan, C. Nolen, D. Teweldebrhan, A. Balandin, *ECS Trans*, **33** (2010)
20. F. Clerc, C. Battaglia, M. Bovet, L. Despont, C. Monney, H. Cercellier, M. Garnier, and P. Aebi, *Physical Review B*, **74**, 155114, (2006)
21. N P Ong, K Maki, *Phys. Rev. B* 32 6582 (1985)
22. ME Itkis, B M Emerling, JW Brill, *Phys. Rev. B* **52,** R11545 (1995)

Mater. Res. Soc. Symp. Proc. Vol. 1344 © 2011 Materials Research Society
DOI: 10.1557/opl.2011.1360

Stable superconducting niobium ultrathin films

Cécile Delacour, Luc Ortega, Bernard Pannetier and Vincent Bouchiat
Institut Néel, CNRS-Université Joseph Fourier-Grenoble INP, BP 166, F-38042 Grenoble, France.

ABSTRACT

We report on a combined structural and electronic analysis of niobium ultrathin films (from 2.5 to 10 nm) epitaxially grown in ultra-high vacuum on atomically flat sapphire wafers. We demonstrate a structural transition in the early stages of Nb growth, which coincides with the onset of a superconducting-metallic transition (SMT). The SMT takes place on a very narrow thickness range (1 ML). The thinnest superconducting sample (3 nm/ 9ML) has an offset critical temperature above 4.2K and allows to be processed by standard nanofabrication techniques to generate air and time stable superconducting nanostructures.

INTRODUCTION

Superconductivity has been recently shown to survive down to extremely confined nanostructures such as metal monolayers [1] or clusters made of few organic molecules [2]. While these structures are extremely interesting to probe the ultimate limits of superconductivity, their studies are limited to in-situ measurements. Preserving a superconducting state in ultrathin films that can be processed by state-of-the-art nanofabrication techniques and withstand multiple cooling cycles remains technologically challenging and is timely for quantum devices applications.

Having the highest critical temperature among elemental superconductors, niobium is an biquitous material for superconducting thin films and its performance is known to increase when epitaxy conditions can be reached. Hetero-epitaxy of niobium is best performed on sapphire substrates since both their lattice parameters and thermal dilatation coefficients match rather well [3]. Most combined structural and transport studies have involved A-plane (1000) oriented sapphire as the growing substrate. However the high step-edge roughness [4] and interfacial stress found for that crystal orientation is likely to be a probable source of rapid aging. Only few studies [5] have described Nb growth on R-plane oriented (1-102) sapphire. However, this orientation provides an ideal substrate to promote epitaxy of Nb with (001) orientation. We present here a combined structural and electronic transport analysis of the early growth stages of niobium films on that orientation with the proof of concept of the interest of these films for quantum device fabrication.

EXPERIMENT DETAILS

The niobium films are grown in ultra-high vacuum using a specific recipe: R-Plane oriented sapphire substrates that are cleaned using water-based surfactants in ultrasonics bath. The surface is then prepared by annealing during 1 hour in air at 1100°C. Wafers are transferred in the deposition chamber (base pressure ~10^{-10} torr). Before niobium deposition, in-situ cleaning of the surface is performed during several hours using Argon milling. Then, the substrate is heated at 660°C and Niobium is deposited using electron gun at a rate of 0.02 nm/s. The vacuum is kept at residual pressures lower than 10^{-9} Torr during niobium growth. Finally, the sample is cooled below 80°C and we covered by evaporation of 2nm thick of silicon in order to protect the film from oxidation.

DISCUSSION

Structural analysis

X-ray diffraction (XRD) measurements reveal that films thicker than 4 nm are single crystalline with a (100) niobium bulk orientation in agreement with the models. The (110) pole figure shows the diffracted intensities of the 4 diagonal (110)-plans of the body center cubic lattice of niobium (χ = 45° +/- 3°) as expected for a hetero-epitaxial growth of (001) on the R-plan sapphire. XRD patterns of samples thinner than 4nm (samples D, E, F) exhibit no trace of the body center cubic Nb lattice, suggesting that hetero-epitaxy is not achieved if the growth is interrupted in its early stages. Further investigations by grazing incidence X-ray diffraction confirm that the thinnest films (samples D, E) are indeed made of mosaic polycrystals, as the patterns are similar to the one encountered for niobium powder (Fig. 1).

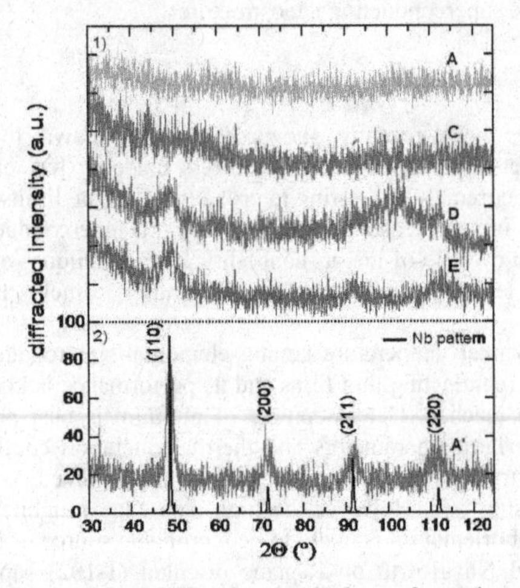

Figure 1. Grazing incidence X-ray diffraction analysis performed on a series of films with different Nb thicknesses : A=9nm, B=8nm, C=4nm, D=3.3nm, E=3nm. Curves are shifted for sake of clarity. To illustrate the difference between hetero-epitaxial (A,C) and polycrystalline films (D,E), the bottom inset show the diffraction pattern of a thicker polycrystalline niobium film (A'= 9 nm) (red curve) and compares it to a theoretical pattern of niobium powder (black curve).

The initial growth stage of niobium film on an atomically flat substrate follows the step-flow model [6] as previously predicted [5]. Niobium adatoms move on the terraces, rearrange themselves along the sapphire step edges forming patches that percolate, leading to a mosaic polycristal film. When the layer get thicker, typically over 10 ML, volumic interactions start to dominate over the step edge 2D interaction and force the film to become single-crystalline with a (100) niobium bulk orientation.

Electronic properties

The square resistance of samples have been measured on strip lines etched from the films using 4 probes techniques from room temperature down to 0.1 K using a dilution refrigerator.

Resistance-temperature curves (Fig. 2) shows superconducting transitions for all films thicker than 3 nm. Critical temperature T_c is found to be inversely proportional with d and the transition widen notably for polycrystalline films (samples D, E). This widening mostly occurs on the onset of the transition as a result of an increase of the superconducting fluctuations. However there is no trace of residual resistance below T_c that would have revealed the existence of phase fluctuations.

The thinnest films (samples F,G) do not show superconducting transition at the lowest temperature (0.1K). For sample F, high resolution measurements have shown a drop at 0.5K towards a partially resistive state which equals 70% of the resistance measured at 10K, suggesting the transition towards an intermediate metallic state. For sample G, this decrease clearly follows a logarithmic dependence with temperature, which is a signature of a 2D weak localization.

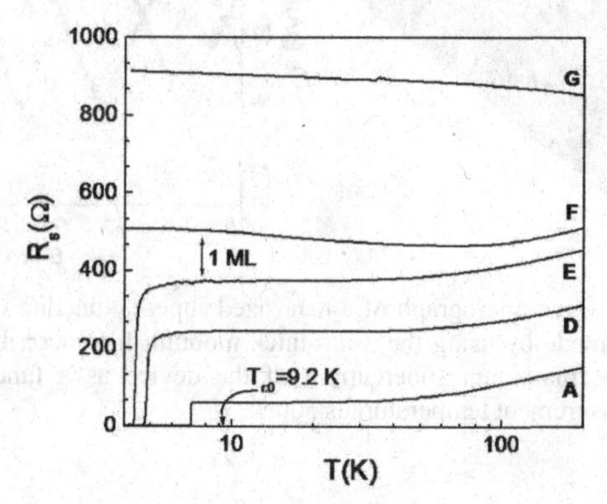

Figure 2. Sheet resistance R_S as a function of temperature. As thickness is reduced (from A=9nm, D=3.3nm, E=3nm, F=2.8nm, G=2.5nm), R_S (10K) increases and the superconducting transition temperature T_C vanishes. An electronic transition occurs for a critical Nb thickness of 2.98 nm (F) on a very narrow thickness range (1 ML).

The suppression of superconductivity has been studied before in several configurations, for granular or amorphous films and with different elemental materials and alloys. Its origins could be multiple as they range from transition to a localized state due to disorder [7] to quantum confinement [8,9] or to inverse proximity effect with the surrounding materials [10]. In a disorder induced superconducting to insulator transition [11], R_s was believed to be of the order of the resistance quantum $h/(4e^2)$=6.5 kΩ In our system however the transition is observed for a well lower resistance per square (Fig. 2) suggesting that the localization is not the dominant factor that leads to the suppression of the superconducting state.

Application to quantum devices

The reported single crystalline niobium films are stable in time over months even exposed to the ambient atmosphere, as demonstrated by the 4 nm Nb-thick with Tc= 5 K that is still preserved after multiples thermal cycling and even 3 years after its growth. The thinnest samples are also resistant to micro and nano-fabrication processes, with a superconducting transition temperature preserved above 4.2K for nanostructures made from sample C. This 3-nm-Nb-thick sample has been patterned using conventional Deep-UV photolithography followed and AFM nanolithography leading to nanometer scaled superconducting quantum interferometer devices (fig.3) [12].

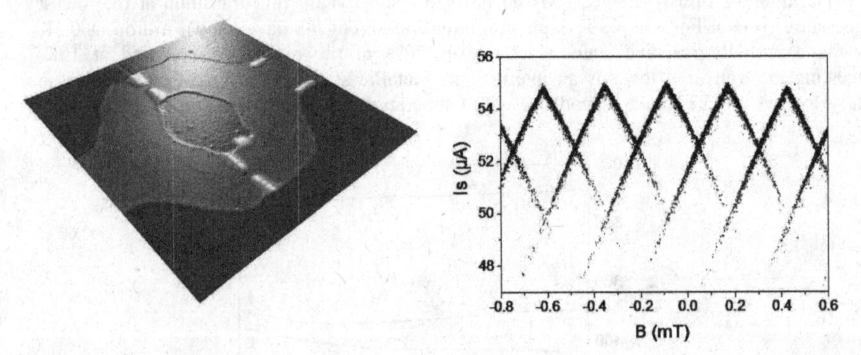

Figure 3. Left: Atomic force micrograph of a nanosized superconducting quantum interference device (SQUID) AFM-made by using the 3-nm-thick niobium film described in the main text. Right: variations of the maximum supercurrent of the device as a function of the applied magnetic field. The measurement temperature is 50mK.

CONCLUSIONS

We have presented a combined analysis of structural and electron transport properties of sub-10 nm niobium films evaporated on R-plane sapphire. The films undergo a sharp non-universal superconducting-metalic transition at 3nm which follow a structural transition from poly to single crystal occurring occurs at thickness slightly above that value. The Tc cut-off occurs over a narrow thickness range (1ML) obtained for a non-universal critical resistance which is more than 10 times below the quantum of resistance $h/(4e2)$. Resistance temperature measurements show a transition toward a metallic state with 2D weak localization features. Above that critical thickness, the overall stability demonstrated by both structural and electronic measurements of the ultrathin films offers a reliable starting material for the realization of quantum superconducting devices.

REFERENCES

1. T. Zhang, et al., Nature Physics **6**, 104 (2010).
2. K. Clark, A. Hassanien, S. Khan, K.-F. Braun, H. Tanaka, and S.-W. Hla, Nature Nanotech **5**, 261 (2010).
3. G. Oya, M. Koishi, and Y. Sawada, Journal of Applied Physics **60**, 1440 (1986).
4. B. Wolfing, K. Theis-Brohl, C. Sutter, and H. Zabel, J. Phys. Condens. Matter **11**, 2669 (1999).
5. A. Wildes, J. Mayer, and K. Theis-Bröhl, Thin Solid Films **401**, 7 (2001).
6. C. Flynn, J. Phys. F: Met. Phys. **18** L195 (1988).
7. A. Finkel'Stein, Physica B: Condensed Matter **197**, 636 (1994).
8. Y. Guo, et al., Science **306**, 1915 (2004).
9. S. Bose, R. Banerjee, and A. Genc, J. Phys. Condens. Matter **18**, 4553 (2006).
10. J. W. P. Hsu, S. I. Park, G. Deutscher, and A. Kapitulnik, Phys. Rev. B **43**, 2648 (1990).
11. A. M. Goldman and N. Markovic, Physics Today **51**, 39 (1998).
12. Bouchiat V., Faucher M., Thirion C., Wernsdorfer W., Fournier T., and Pannetier B., Appl. Phys. Lett. **79**, 123 (2001).

Mater. Res. Soc. Symp. Proc. Vol. 1344 © 2011 Materials Research Society
DOI: 10.1557/opl.2011.1351

Structural and magnetic properties on F-doped $LiVO_2$ with two-dimensional triangular lattice

Yang Li [1,2,*], Xiaoxiang Li [2], Lihua Liu [2], Ning Chen [3], Jose García [1], Rafael Dávila [1], Danny Faica [1], Alfred Rivera [1], Pedro Rodríguez [1], Rubén Pérez [1] and Guohui Cao [2]

[1] Department of Engineering Science and Materials, University of Puerto Rico at Mayaguez, Mayaguez, PR 00681-9044, U.S.A.
[2] Department of physics, University of Science and Technology Beijing, Beijing 100083, China
[3] School of Material Science and Engineering, University of Science and Technology Beijing, Beijing 100083, China

ABSTRACT

The layered oxide $LiVO_2$ recently has received more attention due to its interesting structural and magnetic behaviors involving the two-dimensional magnetic frustration in these systems. We synthesized a series of F-doped $LiVO_2$ samples, and reported the F-doping effect on the structure and transition temperature T_t. The samples $LiVO_{2-x}F_x$ (x=0, 0.1, 0.2 and 0.3) were characterized by X-ray diffraction, scanning electron microscope (SEM), differential scanning calorimetry (DSC), magnetic susceptibility and specific heat measurement. The structural analysis shows that with increasing x, the ratio of lattice parameter c/a increasing, i.e. in the a-b plane the lattice is compressed while in the c-axis direction the lattice expands. The DSC measurements show that a first-order phase transition happens at around 500 K, and the thermal hysteresis around phase transition temperature T_t increases with increasing x. Substitution of O with F ions results in a change of two dimensional characteristics and the distortion of the VO_6 block in structure, which significantly influence the magnetic ordering transition temperature T_t.

INTRODUCTION

There is a great deal of interest in the transition metal compound $LiVO_2$ with two-dimensional triangular lattice, due to it exhibits peculiar structural and magnetic behaviors involving the magnetic frustration. For $LiVO_2$ an orbital ordering transition occurs near 500 K, which leads to a suppression of magnetic moment below this transition temperature T_t [1,2,3]. Magnetic V^{3+} (S=1) ions with partly filled t_{2g} orbitals in this compound occupy the sites of a triangular lattice [4,5,6]. The system forms an orbitally ordered state at low temperatures [2,7,8]. $LiVO_2$ crystallizes into a rhombohedral structure (as shown in Fig. 1) with hexagonal dimensions a = 2.83 and c =14.87 Å [space group $R\bar{3}m$ (No.166)], in which the distribution of ions can be represented by $(Li^+)[V^{3+}]O_2^{2-}$. The V-O coordination is octahedral VO_6 site. The Li, V and O layers stack along c-axis direction. Each of these layers forms a triangular two-dimensional lattice.

$LiVO_2$ exhibits a first order magnetic phase transition [9] at around T_t = 500 K. The interesting property is a change from a high temperature paramagnetic phase with a large negative Curie-Weiss temperature Θ = -1800 K in $\chi \propto 1/(T+\Theta)$, corresponding to a large antiferromagnetic coupling, to a low temperature "nonmagnetic system" without any sign of long-range order [10]. The high temperature susceptibility is consistent with a local moment of S

* Corresponding author: YL electronic mail: ylibp@hotmail.com

= 1, as expected for a d^2 configuration in the presence of a strong on-site Coulomb interaction. Based on the model of d^2 configuration in a *2D* triangular lattice, Pen et al.'s theoretical simulation suggests that this phase transition is driven by a peculiar type of orbital ordering that removes the frustration inherent in the triangular lattice, and $LiVO_2$ undergoes a phase transition into a spin-singlet phase at low temperature [2]. Goodenough [1,11] interpreted the phase transition in terms of the formation of trimers below T_t. In his model, the nonmagnetic behavior is attributed to molecular orbital formation in the basal plane, which would quench the local spin moment. The trimerization model is supported by Tian et al.' electron diffraction observation on $LiVO_2$ single crystals [3]. The superlattice reflections disappear above the phase transition temperature T_t, and reappear below the T_t, which suggests that Goodenough's trimer forms below T_t and vanishes above T_t.

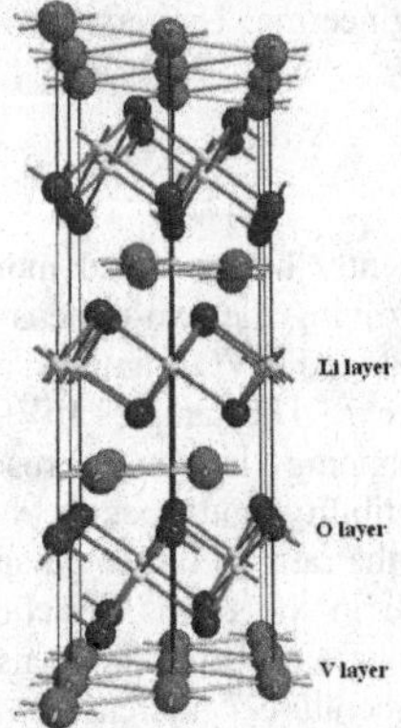

FIG. 1. The structure of $LiVO_2$. The Li, V and O layers stack along *c*-axis direction.

It is significant that the magnetic frustration of $LiVO_2$ heavily depends on microstructure, stoichiometry, and ionic valence. The investigation on the substitution effect on V, Li and O positions is important for understanding magnetic correlation mechanism of V moments. The Cr- and Ti-doped $LiVO_2$ experiments show that Cr and Ti occupy V sites and result in a decrease of phase transition temperature with doping content increasing [1]. It can be explained that Cr and Ti moments suppress magnetic frustration in V planes. Recently we prepared a system of samples $Li_{1-x}Mg_xVO_2$, (x=0, 0.05, 0.10 and 0.15) and investigated the Mg-doping effect on the structural and magnetic properties [12]. With increasing Mg-doping content, the two-dimensional structural characteristic decrease. The resistivity of samples obeys the Mott $T^{-1/4}$ law, which indicates a *3D* variable doping conduction mechanism. Based on the magnetic measurements, the Mg doping undermines the balance of spin-orbital moment within the VO_6 block, which leads to a decrease of phase transition temperature T_t. However, the orbital ordering mechanism of the magnetic phase transition in $LiVO_2$ still remains an open question. To the authors' knowledge, there has not been a general study about the F-doping effect in $LiVO_2$. Unlike Cr- and Ti-dopants on V sites and Mg-dopants on Li sites, the F atoms occupy O sites, which would change structure of VO_6 block and indirectly modify the magnetic frustration structure of V moment. The F-doping will adjust the microstructure and ionic valence of VO_6 block. The investigation on F-doped $LiVO_2$ will be helpful to explain the nature of the magnetic phase transition of $LiVO_2$.

EXPERIMENT

A series of samples of $LiVO_{2-x}F_x$ (x=0, 0.1, 0.2 and 0.3) were prepared by solid state reaction. The chemical reaction is realized as

$$(3-3x)Li_2CO_3+(3-x)V_2O_3+6x\,LiF+2xV\rightarrow 6LiVO_{2-x}F_x+(3-3x)CO_2\uparrow$$

First, LiF, Li_2CO_3 and V_2O_3 were mixed in an agate and sintered at 750 °C for 24 hours. This precursor of Li-F-V-O were ground and mixed fully again, then pressed into a pellet, and annealed at 850 °C for 48 hours followed by furnace cooling. The pellets were under the atmosphere of inert gas flow during the entire annealing process. Samples were analyzed by x-ray powder diffraction. The XRD data refinement is performed using the general structure analysis software (GSAS) package [13,14]. The energy dispersive spectroscopy (EDS) compositional analysis was carried out. The differential scanning calorimetry (DSC) measurements were finished from room temperature to 600 K. The magnetic susceptibility and specific heat were measured in the physical property measurement system (PPMS).

DISCUSSION

1. Structure and composition analysis

Structural analysis for $LiVO_{2-x}F_x$ (x=0, 0.1, 0.2 and 0.3) by powder X-ray diffraction (XRD) showed a characteristic pattern of rhombohedral structure [space group $R\bar{3}m$ (No.166)]. As shown in Fig.2, the XRD pattern and refinement exhibit the x=0.1 sample with single-phase. The lattice parameters of x = 0.1 sample is a = b = 2.839 Å, c=14.821 Å. The XRD refinement shows that the refined compositions of samples are close to the nominal compositions, which also supported by the experiments of EDS analysis.

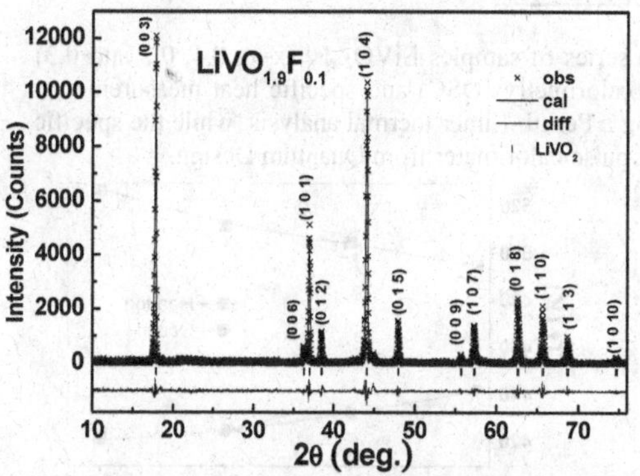

FIG.2. X-ray refinement for $LiVO_{2-x}F_x$ (x=0.1). Upper curve: data and fit, with difference plot below. Ticks show peaks indexed according to rhombohedral structure.

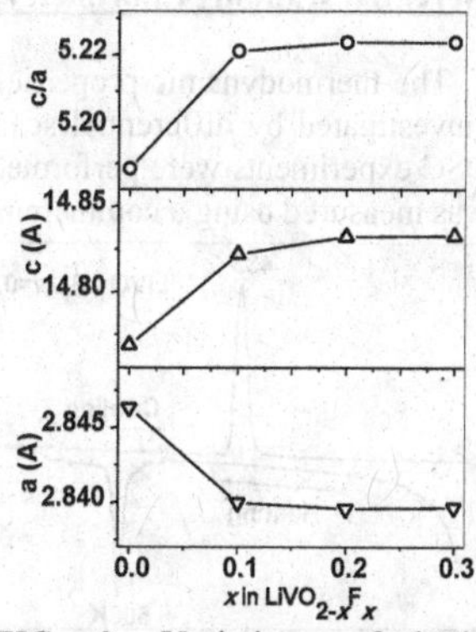

FIG. 3. Variations of lattice parameters and cell volume with x for $LiVO_{2-x}F_x$ (x=0, 0.1, 0.2 and 0.3).

Figure 3 shows the variation of the refined lattice parameters and c/a ratio with increasing x at room temperature. The substitution of smaller F^- ion (r_F = 0.66 Å) for O^{2-} ion (r_O=0.68 Å) causes a variation of lattice parameters. Within the crystallographic a-b plane, the lattice parameter a decreases with increasing x due to the smaller size of F^- ion. However, along c-axis direction, an increase of lattice parameter c and c/a ratio with increasing x suggests some detractions of bonding among Li, O and V ion layers along c-axis., F dopants results in a weaker hybridization in c-axis direction, which enhances the two-dimensional characteristic in $LiVO_2$.

Scanning electron microscopy (SEM) observations were carried out in accordance with standard laboratory practice by using an electron microscope with an energy dispersive spectroscopy (EDS) system for chemical compositional analysis. SEM observations showed that samples are uniform as shown in Fig. 4. There is no texture and the grains are randomly oriented. The average grain size was measured by the linear intercept method to be ~ 1 μm. Chemical compositional analyses of F-doped samples with nominal composition of $LiVO_{2-x}F_x$ (x=0, 0.1, 0.2 and 0.3) were performed on particles selected by the electron diffraction. The quantification results are consistent with the sample nominal compositions.

FIG. 4. The SEM features on section of the sample $LiVO_{2-x}F_x$ (x=0.1).

2. Differential scanning calorimetry and specific heat

The thermodynamic properties of a series of samples $LiVO_{2-x}F_x$ (x=0, 0.1, 0.2 and 0.3) were investigated by differential scanning calorimetry (DSC) and specific heat measurements. The DSC experiments were performed using a Perkin-Elmer thermal analysis, while the specific heat was measured using a commercial heat pulse calorimeter from Quantum Design.

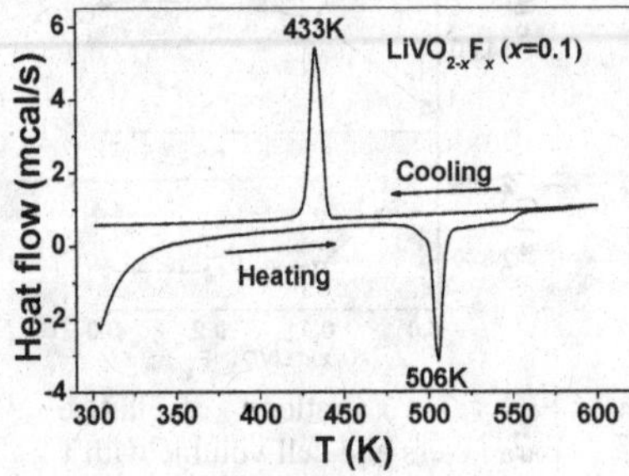

FIG. 5. Phase transition observed, for $LiVO_{2-x}F_x$ (x=0.1) by DSC.

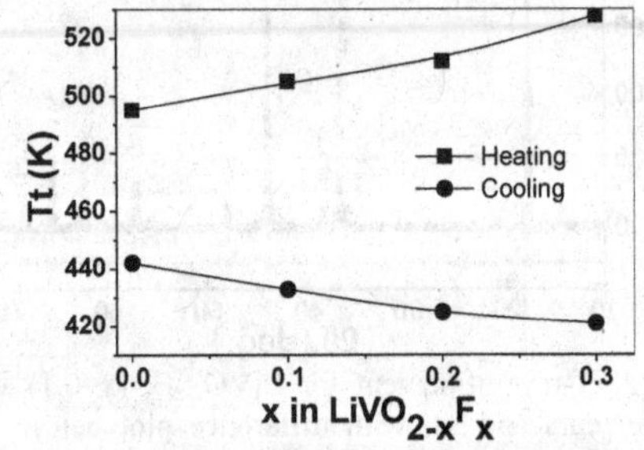

FIG.6. Phase transition temperature T_t with x for $LiVO_{2-x}F_x$ (x=0, 0.1, 0.2 and 0.3).

Fig. 5 shows the DSC of x=0.1 sample traces obtained by using a heating and cooling rate of 10 K/min between 300 and 600 K; it confirmed the first order phase transition at about 506 K on heating and at about 433 K on cooling. The calculated enthalpy change is about 3 kcal/mol.

With increasing concentration of F, around the phase transition, the thermal hysteresis appears significant. As shown in Fig. 6, the phase transition temperature $T_{t,heating}$ on heating process increases with increasing F content, while the phase transition temperature $T_{t,cooling}$ on cooling process decreases with increasing F content. The difference between $T_{t,heating}$ and $T_{t,cooling}$ increases, which is involved with the F-doping induced magnetic structure change.

The temperature dependence of specific heat C was measured from 2 to 300K. As shown in Fig. 8. An upturn observed below 5 K may be due to a Schottky anomaly. In addition, the spin entropy of the presumed V local moments is partially converted to the entropy of the conduction carriers which is manifested as a small (11.7 mJ/mol K^2) zero temperature electronic specific heat coefficient, as shown in Fig. 7. An approximate $C/T = \gamma + \beta T^2$ fit was obtained, with electron and phonon contributions γ = 11.7 mJ/K^2 and β = 0.036 mJ/K^4, respectively. This low temperature behavior of the specific heat can be explained by a spin wave excitation [[15]].

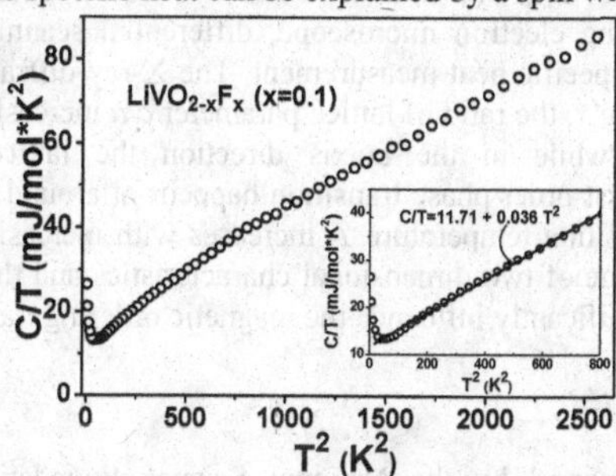

FIG.7. C/T vs. T^2 curve of $LiVO_{2-x}F_x$ (x=0.1). Inset: the C/T with the fitting curve $\gamma+\beta T^2$.

3. Magnetic susceptibility

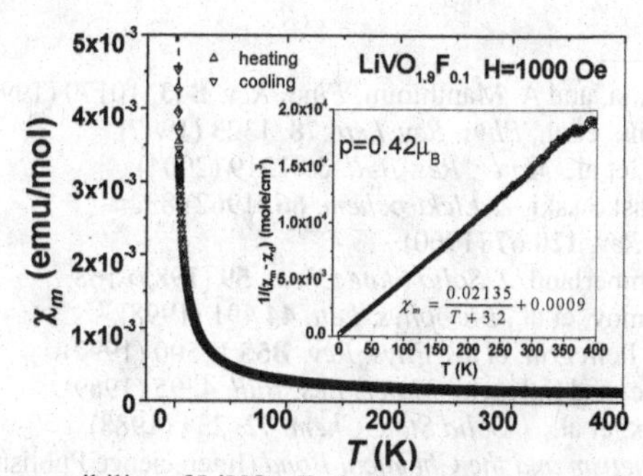

FIG. 8. The dc susceptibility of $LiVO_{2-x}F_x$ (x=0.1) on heating and cooling process.

Measurement of the magnetic susceptibility was performed in an applied field of 1 kOe for the samples $LiVO_{2-x}F_x$ with different F content. The susceptibility exhibits a small value and no thermal hysteresis in heating and cooling process. Based on DSC measurements, the phase transition temperature T_t is higher than 400 K, which is out of the available measurement range of the equipment in our lab. The dc magnetic susceptibility in cooling and heating process for the samples $LiVO_{2-x}F_x$ (x=0.1) is shown in Fig. 8. The susceptibility data can be fitted to the Curie-Weiss law, $\chi_m = [p^2/3k_B(T - \theta_w)+\chi_d]$, to yield the effective moment p_{eff}=0.42 μ_B per V ion, which is much smaller than the value of free-ion V^{3+} moment (p =2.83 μ_B). Moreover, the Weiss temperature is yieled with θ_w= –3.2 K and χ_d=0.0009 emu/mol. There is no magnetic coupling among V moments. Within the available measurement temperature region (T < 400K), we only can anticipate the x = 0.1 sample is in a low temperature "nonmagnetic system" without any sign of long-range order [10]. The susceptibility at higher temperature needs to be measured further for samples in order to investigate magnetic transition temperature with increasing F content.

CONCLUSIONS

A series of samples $LiVO_{2-x}F_x$ (x=0, 0.1, 0.2 and 0.3) were synthesized and characterized by X-ray diffraction, scanning electron microscope, differential scanning calorimetry (DSC), magnetic susceptibility and specific heat measurement. The X-ray diffraction refinement shows that F dopants with increasing x, the ratio of lattice parameter c/a increasing, i.e. in the a-b plane the lattice is compressed while in the c-axis direction the lattice expands. The DSC measurements show that a first-order phase transition happens at around 500 K, and the thermal hysteresis around phase transition temperature T_t increases with increasing x. Substitution of O with F ions results in a change of two dimensional characteristics and the distortion of the VO_6 block in structure, which significantly influence the magnetic ordering transition temperature T_t.

ACKNOWLEDGMENTS

This work was supported by the National Science Foundation (Grant No. DMR-0821284), the National Natural Science Foundation of Beijing (Grant No. 1072007) and NASA (Grant No.NNX10AM80H and NNX07AO30A).

REFERENCES

[1] J. B. Goodenough, G. Dutta, and A. Manthiram, *Phys. Rev.* **B43**, 10170 (1991).
[2] H. F. Pen, J. van den Brink, et al., *Phys. Rev. Lett.*,**78**, 1323 (1997).
[3] W. Tian, M.F. Chisholm, et al., *Mater. Res. Bull.* **39** 1319 (2004).
[4] B. Rutter, R. Weber, J. Jaskowski, *Z. Elektrochem.* **66** (1962) 832.
[5] J.B. Goodenough, *Phys. Rev.* **120** 67 (1960).
[6] T.A. Hewston, B.L. Chamberland, *J. Solid State Chem.* **59** (1985) 168.
[7] S. Yu. Ezhov, V.I. Anisimov, et al., *Europhys. Lett.* **44** 491 (1998).
[8] H.F. Pen, L.H. Tjeng, E. Pellegrin, et al., *Phys. Rev.* **B55** 15500 (1997).
[9] K. Kobayashi, K. Kosuge, and S. Kachi, *Mater. Res. Bull.* **4**, 95 (1969).
[10] L. P. Cardoso, D. E. Cox, et al., *J. Solid State Chem.* **72**, 234 (1988).
[11] J.B. Goodenough, *Magnetism and the Chemical Bond* (Interscience Publishers, New York, 1963).
[12] Yang Li, W.P. Wang, et al., *J. Appl. Phys.* 107, 09E108 (2010).
[13] B. H. Toby, *J. Appl. Cryst.* **34**, 210 (2001).
[14] A. C. Larson and R. B. von Dreele, Tech. Rep. LAUR 86-748, Los Alamos National Lab (2000).
[15] Toshikazu Sato and Yoshihito Miyako, J. Phys. Soc. Jpn. **51** 2143 (1982).

AUTHOR INDEX

SUBJECT INDEX

Printed in the United States
By Bookmasters